Environmental Flows in Water Resources Policies, Plans, and Projects

ENVIRONMENT AND DEVELOPMENT

A fundamental element of sustainable development is environmental sustainability. Hence, this series was created in 2007 to cover current and emerging issues in order to promote debate and broaden the understanding of environmental challenges as integral to achieving equitable and sustained economic growth. The series will draw on analysis and practical experience from across the World Bank and from client countries. The manuscripts chosen for publication will be central to the implementation of the World Bank's Environment Strategy, and relevant to the development community, policy makers, and academia. Topics addressed in this series will include environmental health, natural resources management, strategic environmental assessment, policy instruments, and environmental institutions, among others.

Also in this series:

International Trade and Climate Change: Economic, Legal, and Institutional Perspectives

Poverty and the Environment: Understanding Linkages at the Household Level

Strategic Environmental Assessment for Policies: An Instrument for Good Governance

Environmental Health and Child Survival: Epidemiology, Economics, Experiences

Environmental Flows in Water Resources Policies, Plans, and Projects

Findings and Recommendations

Rafik Hirji and Richard Davis

THE WORLD BANK
Washington, DC

ISBN: 978-0-8213-7940-0
eISBN: 978-0-8213-8012-3
DOI: 10.1596/978-0-8213-7940-0

Cover photo: inner photo, © World Bank; outer photo,
© iStockphoto.com/jameslee999

**Library of Congress Cataloging-in-Publication Data has been
applied for.**

C O N T E N T S

PART III
Case Studies of Environmental Flow Implementation

CHAPTER 4
61 Case Study Assessment

CHAPTER 5
67 Policy Case Studies: Lessons

CHAPTER 6
83 Basin Plan Case Studies: Lessons

CHAPTER 7
93 Project Case Studies: Lessons

PART IV
Mainstreaming Implications

CHAPTER 8
119 Achievements and Challenges

CHAPTER 9
129 Framework for Mainstreaming Environmental Flows

Tables

Foreword

Investments in infrastructure provide opportunities for economic growth and poverty alleviation. Many developing nations face the major development challenge of providing the infrastructure to meet the growing demand for water for domestic consumption, agriculture, energy, and industry and for a buffer against the vulnerabilities to floods and droughts. Climate change is likely to heavily affect water supply and demand and worsen extreme events. Adaptation to climate variability and climate change may require a suite of solutions, including investments in water resources management policies; plans and institutions; demand management; and conservation and protection of watersheds, lakes, wetlands, and aquifers. This adaptation may also require the rehabilitating, upgrading, and constructing new onstream and offstream abstractions, small and large dams, and interbasin transfers, as well as the conjunctive use of surface and ground water.

The global food crisis has refocused attention on improving agriculture, including investment in irrigation infrastructure, among other actions, in developing nations. At the same time, the global energy crisis has drawn attention to accelerating investments in energy production, including hydropower development. The current global financial and economic crisis is adding weight to the argument for increasing investments in infrastructure in the water, transport, energy, and other sectors in developed and developing nations both as a solution to and buffer against the uncertainties associated with the economic downturn. In all cases, the Sustainable Development Network's challenge will be how and at what pace to increase infrastructure investments while maintaining the necessary measures required for economic, social, and environmental sustainability.

The World Bank's 2003 Water Resources Sector Strategy calls for investing in such "high-risk" infrastructure projects as dams in an environmentally and socially responsible manner. It calls for a new business model for developing high-risk water infrastructure that takes full account of both upstream and downstream environmental and social impacts of the infrastructure in a timely, predictable, and cost-effective manner. Apart from reducing uncertainties associated with project decision making and financing, this socially and environmentally responsible approach will help sustain ecosystem services on which many poor people in developing countries rely. The formation of the Sustainable Development Network in 2007 has further elevated environmental responsibility as a core element of World Bank's work.

The World Bank's own analysis and the far-reaching report of the World Commission on Dams have both shown that dam developments have not always been planned, designed, or operated satisfactorily. Even though dams generate considerable benefits in aggregate, these benefits have not always been shared equitably. Dams have often been developed without adequate consideration for either the environment or the people downstream of the dam who rely on local ecosystem-based services.

The World Bank's knowledge of, and experience in, addressing the upstream impacts of dams has advanced considerably over recent decades. However, its experience in addressing the downstream impacts of water resources infrastructure, although growing, remains limited. Environmental flow work within the Bank has been shaped by the evolving global knowledge, practice, and implementation of environmental flows. The Bank has also contributed to this growing international experience, particularly through its support for the Lesotho Highland Water Project, the restoration of the downstream parts of the Tarim River, and the restoration of the Northern Aral Sea and the Senegal River basin. It has also supported environmental flow initiatives in Central Asia, China, Ecuador, India, Mexico, the Mekong River region, Moldova, Tajikistan, Tanzania, and Ukraine, and has produced knowledge products and support material, including a series of technical notes on environmental flows.

This report further contributes to international knowledge about environmental flows and sustainable development. It focuses on the integration of environmental water allocation into integrated water resources management (IWRM) and so fills a major gap in knowledge on IWRM. It also contributes to broadening our understanding of benefit sharing from risky infrastructure development. This report is an output of an important collaboration between the Bank's Environment Department and Energy, Transport, and Water Department to promote and mainstream sustainable development.

James Warren Evans
Director
Environment Department

About the Authors

Rafik Hirji, a senior water resources specialist at the World Bank, has extensive experience in water resources planning, management, and development projects and programs in Africa, Asia, the Caribbean, Yemen, and the United States. He has led the Sustainable Development Network's agenda on water and environment by promoting operational tools for sustainable utilization and management of rivers, lakes, and aquifers. He was team leader for the global lake basin management initiative, sector analysis on strategic environmental assessments in water resources management, and sector analysis for the integration of environmental flows into water resources operations and water policy dialogue. He has also led water resources policy dialogue and the preparations of national water resources strategies for Kenya, Tanzania, and Trinidad and Tobago and supported the preparation of state water plans in Tamil Nadu and Orissa, India, and the Ghana Water Resources Management Study. Currently, Dr. Hirji is leading the preparations of the global groundwater governance project and supporting the World Bank's flagship work on climate change and water, including managing the preparation of two special papers on impacts on groundwater resources and adaptation options and climate change and adaptation implications on freshwater ecosystems. Dr. Hirji has published widely and is the principal editor for the SADC regional report on environmentally sustainable water resources management in southern Africa and co-editor for the World Bank Water Resources and Environment Technical Note series. He holds a MSc in environmental engineering and science and a PhD in water resources planning from Stanford University. He is a registered professional engineer in the U.S.

Richard Davis, a senior science advisor to the Australian National Water Commission, has had an extensive career in water and environmental research with CSIRO, Australia, specializing in environmental flows, water quality, catchment management, and decision support systems. He has also worked for Australian government policy departments and as program coordinator for Land and Water Australia. Dr. Davis was seconded to the World Bank's Environment Department in 2001-03, and he has since consulted widely on country water resources assistance strategies, river and lake basin management operations, environmental flows, and strategic environmental assessment sector analysis. He has published extensively. He was the principal editor of the World Bank Water Resources and Environment Technical Note series. Dr. Davis has a BSc from Otago University, New Zealand, and a PhD and a BEc from the Australian National University.

Acknowledgments

Environmental Flows in Water Resources Policies, Plans, and Projects: Findings and Recommendations was prepared by Rafik Hirji (ETWWA) and Richard Davis (consultant). It, and a complementary report that summarizes 17 case studies of water policies, river basin/catchment plans, and infrastructure development and rehabilitation projects, is based on the economic and sector analysis (ESW)—Mainstreaming Environmental Flow Requirements into Water Resources Investments and Policy Reforms—that was jointly supported by the Environment Department and Energy, Transport, and Water Department and completed in June 2008. The authors are grateful for the support they received from individuals within and outside the World Bank. Editorial support was provided by Robert Livernash and Elizabeth Forsyth. The preparation of this report was funded by the World Bank–Netherlands Water Partnership Program (BNWPP) Trust Fund.

The economic and sector analysis was prepared through an extensive collaboration over 2 years with over 75 water and environment experts, including task team leaders and project staff, researchers, and practitioners from the World Bank and other institutions in many parts of the world. The core team comprised Rafik Hirji (team leader), Richard Davis (consultant), Kisa Mfalila (consultant), and Marcus Wishart (AFTWR). Michelle De Nevers, Laura Tlaiye, Abel Mejia, James Warren Evans, and Jamal Saghir provided overall guidance. Daryl Fields provided detailed comments on an earlier draft. Stephen Lintner provided extensive critiques and comments on earlier drafts.

Case studies 2 and 16 were drafted by Mike Acreman (consultant, United Kingdom); case study 12 by Denise Dalmer (consultant, Canada); case study 11 by Marcus Wishart (World Bank); and case study 7 by Kisa Mfalila (consultant). Summaries of agency and nongovernmental organization practices were provided by Karin Krchnak (the Nature Conservancy), Gregory Thomas (NHI), Kisa Mfalila (WWF, UNDP, UNEP), and Mike Acreman (IUCN, IWMI).

The authors wish to acknowledge the following staff and colleagues who commented on the 17 case studies and provided information and materials: Masood Ahmad, Greg Browder, Ousmane Dione, Jane Kibbassa, Andrew Macoun,

Doug Olson, Geoff Spencer, and Mei Xie of the World Bank; and Mike Acreman (consultant, United Kingdom), Fadhila Hemed (National Environment Management Council, Tanzania), Harry Biggs (SANParks, South Africa), Cate Brown (Southern Waters, South Africa), Satish Choy (Queensland Department of Natural Resources and Water, Australia), Kevin Conlin (BC Hydro, Canada), Mark Dent (University of KwaZulu-Natal South Africa), Saidi Faraji (Ministry of Water and Irrigation, Tanzania), A. J. D. Ferguson (consultant, United Kingdom), Sue Foster (BC Hydro, Canada), Dana Grobler (Blue Science Consulting, South Africa), Larry Haas (consultant, United Kingdom), Thomas Gyedu-Ababio (SANParks, South Africa), Robyn Johnston (Murray-Darling Basin Commission, Australia), Sylvand Kamugisha (IUCN, Tanzania), David Keyser (Trans-Caledon Tunnel Authority, South Africa), Jackie King (University of Cape Town, South Africa), Josephine Lemoyane (IUCN, Tanzania), Delana Louw (Water for Africa consultant, South Africa), John Metzger (consultant, Mekong River Commission), Willie Mwaruvanda (Rufiji Basin Water Office, Ministry of Water and Irrigation, Tanzania), Bill Newmark (Utah Museum of Natural History, United States), Tally Palmer (University of Technology, Sydney, Australia), Sharon Pollard (Association for Water and Rural Development, South Africa), Donal O'Leary (Transparency International, United States), Geordie Ratcliffe (Freshwater Consulting Group, South Africa), Paul Roberts (formerly with the Department of Water Affairs and Forestry, South Africa), Kevin Rogers (University of Witwatersrand, South Africa), Nigel Rossouw (Trans-Caledon Tunnel Authority, South Africa), Hamza Sadiki (Pagani Basin Water Office, Ministry of Water and Irrigation, Tanzania), Charles Sellick (Charles Sellick & Associates, South Africa), Doug Shaw (the Nature Conservancy, Florida, United States), Tente Tente (Trans-Caledon Tunnel Authority, South Africa), Malcolm Thompson (Department of Environment, Water, Heritage, and the Arts, Australia), Pierre de Villiers (Blue Science Consulting, South Africa), Niel van Wyk (Department of Water Affairs and Forestry, South Africa), Bill Young (Commonwealth Scientific and Industrial Research Organisation, Australia), and Bertrand van Zyl (Department of Water Affairs and Forestry, South Africa).

The authors are especially indebted to Washington Mutayoba (Ministry of Water and Irrigation, Tanzania) and Barbara Weston (Department of Water Affairs and Forestry, South Africa) for facilitating reviews of three case studies from Tanzania and three case studies from South Africa by various staff and professional colleagues from their respective countries, and to Steve Mitchell (Water Research Commission, South Africa) for his encouragement and for providing access to research reports from South Africa. The review of the BNWPP environmental flow work drew from an earlier report coauthored by Thomas Panella (now at the Asian Development Bank).

World Bank peer reviewers were Claudia Sadoff, Salman Salman, and Juan D. Quintero. External peer reviewers were Brian Richter (the Nature Conservancy) and John Scanlon (UNEP). Comments were also received from Vahid Alavian, Julia Bucknall, Usaid El-Hanbali, Stephen Lintner, Christine Little, Glenn Morgan, Grant Milne, Abel Mejia, Doug Olson, Stefano Pagiola, Salman Salman, Geoff Spencer, and Peter Watson (former director of infrastructure from the Africa Region).

Abbreviations

BBM	building block method
BNWPP	Bank–Netherlands Water Partnership Program
BP	Bank procedures (World Bank)
CAS	country assistance strategy (World Bank)
CEA	country environmental assessment
COAG	Council of Australian Governments
CWRAS	country water resources assistance strategy (World Bank)
DANIDA	Danish International Development Agency
DPL	development policy lending (World Bank)
DRIFT	downstream response to imposed flow transformation
EFA	environmental flow assessment
EIA	environmental impact assessment
ESW	economic and sector work (World Bank)
EU	European Union
GEF	Global Environment Facility
GLOWS	Global Water for Sustainability (USAID)
IFIM	instream flow incremental methodology
IFR	instream flow requirement
IUCN	International Union for the Conservation of Nature
IWMI	International Water Management Institute
IWRM	integrated water resources management
LHDA	Lesotho Highlands Development Authority
LHWP	Lesotho Highlands Water Project
LKEMP	Lower Kihansi Environmental Management Project
MDG	Millennium Development Goals
NGO	nongovernmental organization
NHI	Natural Heritage Institute
NWI	National Water Initiative (Australia)
OKACOM	Okavango River Basin Water Commission
OMVS	Organisation pour la Mise en Valeur du Fleuve Sénégal
OP	operational policy (World Bank)
PAD	project appraisal document (World Bank)
SAR	staff appraisal document (World Bank)
SDN	Sustainable Development Network (World Bank)
SEA	strategic environmental assessment

TNC	The Nature Conservancy
UNDP	United Nations Development Programme
UNEP	United Nations Environment Programme
UNESCO	United Nations Education, Scientific, and Cultural Organization
USACE	U.S. Army Corps of Engineers
USAID	U.S. Agency for International Development
WANI	Water and Nature Initiative (IUCN)
WFD	Water Framework Directive (European Union)
WRMP	water resources management policy (World Bank)
WRSS	water resources sector strategy (World Bank)
WWF	World Wide Fund for Nature

Notes: Unless otherwise noted, all dollars are U.S. dollars.
All tons are metric tons.

Overview

ENVIRONMENTAL FLOWS ARE REALLY about the equitable distribution of and access to water and services provided by aquatic ecosystems. They refer to the quality, quantity, and timing of water flows required to maintain the components, functions, processes, and resilience of aquatic ecosystems that provide goods and services to people.

Environmental flows are central to supporting sustainable development, sharing benefits, and addressing poverty alleviation. Yet allocating water for environmental uses remains a highly contested process. Investments in water resources infrastructure, especially dams for storage, flood control, or regulation, have been essential for economic development (including hydropower generation, food security and irrigation, industrial and urban water supply, and flood and drought mitigation), but, when they are improperly planned, designed, or operated, they can cause problems for downstream ecosystems and communities because of their impact on the volume, pattern, and quality of flow. While aquatic life depends on both the quantity and quality of water, changes in flows are of particular concern because they govern so many ecosystem processes. Consequently, changes in flow have led to a diminution of the downstream ecosystem services that many of the poorest communities rely on for their livelihoods. In order to achieve sustainable development, downstream impacts will require more attention by all parties, as countries—through both public and private sector investments—expand their infrastructure in many sectors, especially dams for various purposes.

Climate change is projected to affect the supply of and demand for water resources; in turn, these changes will have an impact on water for the environment. Sea-level rise will cause saltwater intrusion and affect estuarine processes that rely on freshwater environmental flows. In some nations, adaptation to climate change is likely to involve more investment in dams and reservoirs to buffer against increased variability in rainfall and runoff. This will further affect downstream ecosystems, unless the impacts are properly assessed and managed.

The overall goal of the analysis presented in this report is to *advance the understanding and integration in operational terms of environmental water allocation into integrated water resources management.* The specific objectives of this report are the following:

- Document the changing understanding of environmental flows, by both water resources practitioners and by environmental experts within the Bank and in borrowing countries
- Draw lessons from experience in implementing environmental flows by the Bank, other international development organizations with experience in this area, and a small number of developed and developing countries
- Develop an analytical framework to support more effective integration of environmental flow considerations for informing and guiding (a) the planning, design, and operations decision making of water resources infrastructure projects; (b) the legal, policy, institutional, and capacity development related to environmental flows; and (c) restoration programs
- Provide recommendations for improvements in technical guidance to better incorporate environmental flow considerations into the preparation and implementation of lending operations.

Environmental Flows: Science, Decision Making, and Development Assistance

The provision of flows, including volumes and timings, to maintain downstream aquatic ecosystems and provide services to dependent communities has been recognized in developed countries for more than two decades and is increasingly being adopted in developing countries. These services include the following:

- Clean drinking water
- Groundwater recharge
- Food sources such as fish and invertebrates
- Opportunities for harvesting fuelwood, grazing, and cropping on riverine corridors and floodplains
- Biodiversity conservation (including protection of natural habitats, protected areas, and national parks)

- Flood protection
- Navigation routes
- Removal of wastes through biogeochemical processes
- Recreational opportunities
- Cultural, aesthetic, and religious benefits.

But the impacts of development on communities downstream are often diffuse, long term, poorly understood, and inadequately addressed.

Assigning water between environmental flows and consumptive and nonconsumptive purposes is a social, not just a technical, decision. However, to achieve equitable and sustainable outcomes, these decisions should be informed by scientific information and analysis. The causes of changes in river flow can also be broader than just the abstraction or storage of water and the regulation of flow by infrastructure; upstream land-use changes due to forestry, agriculture, and urbanization can also significantly affect flows. The impacts of environmental flow can extend beyond rivers to groundwater, estuaries, and even coastal areas.

Many methods, from the very simple to the very complex, exist for estimating environmental flow requirements. The process for estimating environmental flow requirements is also referred to as environmental flow assessment (EFA). There is an extensive body of experience for the main EFA techniques.

The Entry Points for Bank Involvement

The Bank has four entry points through which to support countries seeking to integrate environmental flows into their decision making: (1) water resources policy, legislation, and institutional reforms;[1] (2) river basin and watershed planning and management;[2] (3) investments in new infrastructure; and (4) rehabilitation or reoperation of existing infrastructure or restoration of degraded ecosystems. Consistent with its commitment to sustainable development, the Bank should support measures to promote the integration of environmental flows at an early stage in the decision-making process through dialogue on water resources policy, river basin planning, and programs that entail major changes in land-use. The World Bank already has supported some projects with successful environmental flow components and outcomes.

Environmental Flows, Integrated Water Resources Management, and Environmental Assessment

EFAs are an intrinsic part of integrated water resources management. Although it is desirable for EFAs to be integrated into strategic environmental assessments (SEAs) for policy, plan, program, or sectorwide lending, and into environmental impact assessments (EIAs) for project-level investments, the practice of SEA and EIA has yet to mature to the point at which it can effectively integrate EFA. As a consequence, most EFAs have been undertaken separately either in conjunction with or after the EIAs have been completed.

Bank Adoption of Environmental Flows

An analysis of select dam projects found that, until the mid-1990s, Bank support for environmental and social work was heavily focused on evaluating and addressing the upstream impacts of dams. By the mid-1990s, these assessments had expanded to include downstream environmental and social issues with about equal frequency, underscoring the evolving concern about downstream impacts. An analysis of country water resources assistance strategies, however, showed mixed results concerning the inclusion of environmental flows, with only some countries incorporating them into their planning. There is a limited perception of the need for including environmental flows within the water policies of developing countries, but a good understanding of the importance of environmental flows in catchment-scale water resources planning. The Bank–Netherlands Water Partnership Program has catalyzed some notable achievements in introducing environmental flows into infrastructure planning, design, and operations in dam rehabilitation and reoperation projects.

International Development Organizations and NGOs

Various international development organizations and nongovernmental organizations (NGOs) have been supporting environmental flow assessments at both the project and basin levels, conducting training courses, and providing information and support material. The Bank has partnered with some of these organizations to produce analytical material on the incorporation of environmental flows into infrastructure development and reoperations.

Environmental Flow Implementation Case Studies

Seventeen case studies were selected for an in-depth analysis to identify the lessons from incorporating environmental flows into water resources policy, basin and catchment plans, new infrastructure projects, and the rehabilitation and reoperation of existing infrastructure (Hirji and Davis 2009a). The analysis included eight case studies that were supported by the World Bank.

The assessment criteria included factors that influenced the case study's success, as well as the institutional drivers that initiated and supported the introduction of environmental flows.

Inclusion of Environmental Flows in Water Resources Policies

An analysis of five policy case studies found that the inclusion of environmental flows in policy should provide for the following:

- Legal standing for environmental water allocations
- Inclusion of environmental water provisions in basin water resources plans
- Assessment of all relevant parts of the water cycle when undertaking EFAs
- A method or methods for setting environmental objectives in basin plans

- Attention to both recovery of overallocated systems and protection of unstressed systems
- Clear requirements for stakeholder involvement
- An independent authority to audit implementation
- A mechanism for turning value-laden terms into operational procedures.

Inclusion of Environmental Flows in Basin and Catchment Plans

Several lessons emerged from the analysis of four basin and catchment water resources plans:

- Recognition of environmental flows in water resources policy and legislation provides important backing for including environmental flows in basin or catchment plans.
- There is a need to demonstrate the benefits from environmental water allocations after plans are implemented.
- The term "environmental flows" can be counterproductive if not explained at an early stage.
- Participatory methods need to be tailored to suit stakeholder capacity.
- A range of EFA techniques is needed to suit different circumstances.
- Ecological monitoring is essential to provide information for adaptive management.

Inclusion of Environmental Flows in Infrastructure Projects

Four new dams and four restoration projects were reviewed for lessons in assessing and implementing environmental flows:

- Engineering improvements usually have to be combined with reoperations to provide the volume of water needed for major ecosystem restoration.
- Inclusion of environmental flows in water resources policy simplifies the application of EFAs at the project level.
- Environmental outcomes need to be linked closely to social and economic outcomes.
- EFAs should be conducted for all components of the hydrological cycle.
- Traditionally trained water resources professionals can find it difficult to grasp environmental flow concepts.
- Water resources plans provide benchmarks for water allocations during project assessments.
- Active monitoring is needed to enforce flow allocation decisions and undertake adaptive management.
- It is important to present information in terms that are comprehensible to decision makers.
- Economic studies can support arguments for downstream water allocations.
- EFAs are yet to be fully mainstreamed into EIAs.

- The cost of conducting EFAs constitutes a small fraction of project costs.
- EIAs have not always or adequately identified issues associated with downstream water provisions.

Mainstreaming Implications

The science underpinning EFAs has advanced considerably. There are now many more methods for estimating environmental flow requirements, and more information is available on the ecological response to different flow regimes. There is also growing experience in integrating information from across a range of physical, ecological, and socioeconomic disciplines. In addition, a wide variety of EFA methods have been developed, backed by considerable field experience, to suit a variety of levels of environmental risk, time and budget constraints, and levels of data and skills. The Bank's support for the Lesotho Highlands Water Project has contributed to the development of a method known as Downstream Response to Imposed Flow Transformation (DRIFT), which systematically addresses the downstream biophysical and socioeconomic impacts. There is also a growing body of experience in implementing environmental flows, including monitoring and adaptation of management procedures.

Mainstreaming Achievements

Developed countries, including parts of the United States, Australia, New Zealand, and the countries of the European Union, together with South Africa, have accepted the need to develop and implement catchment water resources plans that include environmental flows. There is general public acceptance of the importance of maintaining healthy aquatic environments in these countries. In these countries, where environmental flows have now been mainstreamed into water resources planning, there is an acceptance that the concept of environmental flows should be extended to groundwater as well as to estuaries and even near-shore regions.

Support for Developing Countries

International development organizations, NGOs, and research organizations have been active in providing support in developing countries through assistance with EFA and implementation, training programs, and provision of support material and Internet resources. The Bank has collaborated with diverse development partners. The Bank's major contribution to global good practice has been its restoration of the degraded Tarim basin and Northern Aral Sea, its assistance with the provision of flood flows in the Senegal basin, its support for the pioneering work on the Lesotho Highlands Water Project, and its growing influence in introducing environmental flows into government water policies. In these cases, provision of environmental flows has restored (or retained) ecosystems with demonstrable benefits to downstream populations; in the Tarim basin case, there were also significant benefits to the upstream irrigation communities.

Challenges

Both the Bank and environmental flows practitioners face many challenges:

- Overcoming the misperceptions arising from the term "environmental flows"
- Developing methods for systematically linking biophysical and socioeconomic impacts
- Incorporating the whole water cycle (surface, groundwater, and estuaries) into the assessments
- Applying EFAs to land-use activities that intercept and exacerbate overland flows
- Including climate change in the assessments
- Integrating environmental flow assessments into strategic, sectoral, and project EAs
- Understanding the circumstances in which benefit sharing is a viable approach.

Framework for Expanded Bank Engagement with Environmental Flows

The analysis points to a four-part framework for improving the Bank's use of environmental flows.

First, efforts are needed to strengthen Bank capacity in assessing and overseeing environmental flows:

- Promote the development of a common understanding across the water and environmental communities about the concepts, methods, and good practices related to environmental flows, including the need to incorporate EFAs into environmental assessment at both project (EIAs) and strategic (SEAs) levels.
- Build the Bank's in-house capacity in EFA by broadening the pool of ecologists, social scientists, and environmental and water specialists trained in EFA.

Second, efforts are needed to strengthen environmental flow assessments in lending operations through training, support materials, and access to international experts:

- Disseminate existing guidance material concerning the use of EFAs in program and project settings and conduct training for Bank and borrower country staff on this emerging issue.
- Identify settings, approaches, and methods for the select application of EFAs in the preparation and implementation of project-level feasibility studies and as part of the planning and supervisory process.
- Provide support for hydrological monitoring networks and hydrological modeling to provide the basic information for undertaking EFAs.
- Prepare an update of the EA sourcebook concerning the use of EFAs in SEAs and EIAs.

- Prepare a technical note that defines a methodology for addressing downstream social impacts of water resources infrastructure projects.
- Test the application of EFAs to include infrastructure other than dams that can affect river flows, as well as other activities, such as investments in large-scale land-use change and watershed management, and their effects on downstream flows and ecosystem services.
- Broaden the concept of environmental flows for appropriate pilot projects to include all affected downstream ecosystems, including groundwater systems, lakes, estuaries, and coastal regions.
- Develop support material for Bank staff and counterparts in borrowing countries, such as case studies, training material, technical notes, and analyses of effectiveness.

Third, efforts are needed to promote the integration of environmental flows into policies and plans through dialogue, instruments such as country water resources assistance strategies (CWRASs), country assistance strategies (CASs), country environmental assessments, and development policy lending, and support material for Bank staff:

- Promote basin or catchment plans that include environmental flow allocations, where relevant, through country dialogue.
- Use CASs and CWRASs to promote Bank assistance with basin or catchment planning and water policy reform so that the benefits of environmental water allocations for poverty alleviation and the achievement of the Millennium Development Goals are integrated into country assistance.
- Incorporate environmental water needs into Bank SEAs such as country environmental assessments and sectoral environmental assessments.
- Test the use of EFAs in a small sample of sectoral adjustment lending operations, including where the sectoral changes will lead to large-scale land-use conversion.
- Promote the harmonization of sectoral policies with the concept of environmental flows in developing countries and the understanding of sectoral institutions about the importance of considering the impact of their policies on downstream communities.
- Develop support material for Bank staff on the inclusion of environmental flows into basin and catchment planning and into water resources policy and legislative reforms.
- Draw lessons from developed countries that have experience with incorporating environmental flows in catchment planning.

Fourth, efforts are needed to expand collaborative partnerships:

- Expand collaboration with NGOs (International Union for the Conservation of Nature, Worldwide Fund for Nature, the Nature Conservancy, Natural Heritage Institute, and others), research organizations, and international organizations

(United Nations Environment Programme, Ramsar Secretariat, International Water Management Institute, and United Nations Education, Scientific, and Cultural Organization) to take advantage of their experience in conducting EFAs and building environmental flow capacity in developing countries.

- Strengthen collaborative relationships with industry associations, such as International Hydropower Association and private sector financing, to extend their recognition of environmental flows as desirable hydrological outcomes to include the social and economic outcomes that result from the ecosystem services delivered by the downstream flows.
- Integrate lessons from the ESW into—and coordinate the activities outlined above with—the ongoing initiative of the World Bank's Sustainable Development Network and Energy, Transport, and Water Department for enhancing benefits to local communities from hydropower projects.

Adoption of this framework will improve the Bank's ability to implement its strategy of increasing investment in water resources infrastructure, while reducing the risk of detrimental environmental impacts that threaten the livelihoods of downstream communities.

Notes

1 The word "policy" is used throughout much of the report to include legislation supporting the policy.

2 Different countries use different terminology: river basins, catchments, and watersheds. Generally river basins are larger than catchments and watersheds. In this report we use the term basin to refer to basins, catchments, and watersheds generically unless a particular catchment or watershed is being discussed.

PART I

Context and Rationale

CHAPTER 1

Introduction

ENVIRONMENTAL FLOWS ARE CONCERNED with the equitable sharing and sustainable use of water resources. They form a central, yet underappreciated and inadequately addressed, element of integrated water resources management (IWRM) (Hirji and Davis 2009b). They lie at the center of the development debate on environmentally responsible water resources development. They also form an integral part of adaptation responses and strategies for addressing climate change.

The debate about environmental flows is really a debate about equity. It is a debate about the allocation of water needed for immediate consumption, often through development investments, and water needed to sustain ecosystem services on which communities and biological diversity have traditionally depended. It is a debate about (a) recognizing that there is a physical limit beyond which a water resource suffers irreversible damage to its ecosystem functions and (b) systematically balancing the multiple water needs of society in a transparent and informed manner.

Environmental flows are the flow regimes needed to maintain important aquatic ecosystem services. They are a core element of good practice in water resources planning and management. While there are numerous definitions of environmental flows, they are defined here as "the quality, quantity, and timing of water flows required to maintain the components, functions, processes, and resilience of aquatic ecosystems which provide goods and services to people" (Nature Conservancy 2006). In some countries, they are regarded as a luxury of a few environmentally conscious people at the expense of scarce water for production needed by many. This misperception has arisen largely because the term "environmental flows" conjures up images of water being allocated to the environment at the expense of human use and economic development or being wasted by being allowed to flow to the sea.[1] The reality is that, rather than being at the expense of people, environmental flows are essential for providing both direct and indirect benefits on which current and future generations rely (see appendix A).[2]

The flow in rivers varies throughout a year and between years. This pattern of flow—termed the flow regime—typically consists of low flows during the drier months, small peaks (freshets) when rains return, and occasional high floods in unregulated rivers. Groundwater levels also can vary naturally throughout a year and between years in response to changes in recharge and discharge. An environmental flow assessment (EFA) is a process used to understand and define the ecosystem functions supported by the various components of flow in a river or groundwater system.

While the EFA is a technical scientific process that links flow regimes and levels with ecosystem outcomes, the allocation of water between the environment's needs and consumptive needs is a societal decision that is undertaken in a multisectoral decision framework. Thus the allocation of environmental flow is a distinctive element of IWRM, a framework that many developed and developing countries are gradually embracing. Box 1.1 highlights the key linkages between environmental flows and IWRM.

BOX 1.1
Environmental Flows and IWRM Linkages

The environment is linked to IWRM in three fundamental ways. First, the aquatic (and related terrestrial) ecosystem provides habitat for fish, invertebrates, and other fauna and flora. The aquatic ecosystem is thus a water-consuming sector just like agriculture, energy, and domestic and industrial supply. Second, the design and operation of hydraulic infrastructure for water supply, sewerage, irrigation, hydropower, and flood control often affect ecosystems, both upstream and downstream of the infrastructure, and communities—farming, pastoral, and fishing—dependent on those ecosystems. Conversely, the reoperation and rehabilitation of existing infrastructure have been used to support the successful restoration of degraded riverine ecosystems. Third, integrated water resources planning and management are facilitated by policies, laws, strategies, and plans that are multisectoral, based on the allocation of water for all uses; protection of water quality and control of pollution; protection and restoration of lake basins, watersheds, groundwater aquifers, and wetlands; and control and management of invasive species.

Contested debates about environmental flows have often arisen when major infrastructure projects, especially dams and direct abstractions, are being planned, designed, constructed, or operated. The development benefits of dams—hydropower generation, water supply, irrigation, regulation of flood control, and abstractions—are usually well quantified and apparent to decision makers. Their detrimental impacts on upstream communities affected by reservoir inundation[3] now receive concerted attention through resettlement programs and action plans. But the impacts of dam development on those downstream of the dam are often diffuse, long term, poorly understood, and inadequately addressed. The downstream impacts—biophysical and social—arise primarily from changes in the quantity, timing, and quality of the flow pattern of rivers. They typically include the following:

- Reduced abundance of fish and invertebrates such as prawns and shellfish
- Reductions in floodplain sediment and nutrient deposition
- Reductions in areas available for floodplain grazing, cropping, and fuelwoods
- Impediments to riverine navigation and transport
- Reductions in water to terrestrial habitats (including protected areas) and aquatic habitats important for biodiversity
- More difficult access to domestic, irrigation, and livestock water supplies
- Changes in estuarine productivity from altered flow patterns and saline intrusion
- Reductions in groundwater recharge
- Loss of cultural amenities.

Downstream communities can often be affected in two fundamental ways when a dam is developed. First, their livelihoods can be disrupted by the changes in the river flow regime from the development itself. Second, the benefits of the development (for example, electricity generated) often end up in distant places, such as urban areas, and the local communities rarely share in those benefits. Programs for sharing benefits from water infrastructure projects could both address downstream impacts and integrate environmental flows into water resources decision making.

Although most attention in the current debate on environmental flows has been directed to the effects of dams and other water resources infrastructure on downstream flows, other development activities, particularly large-scale land-use changes, can also affect the access to water of people downstream of the development. For example, conversion of land into farmland, urban development, or forestry plantations in the headwaters of catchments can accelerate or intercept runoff. These activities can also increase erosion and exacerbate sediment loads and transport. Yet these activities have seldom been considered as requiring environmental flow assessments, even though they can cause significant reductions in downstream river flows and alter river morphology and ecosystem functions.

Climate change is likely to make environmental flows both more important and more difficult to maintain. The annual average inflows of water to surface water and recharge to groundwater systems will be affected by climate change, with consequent impacts on aquatic ecosystems and the ecosystem services that they provide. The frequency of extreme events will also be affected by global warming, causing changes in the frequency of floods and droughts, on which some riverine ecosystems rely. The rise in sea level will affect freshwater inflows into estuarine and coastal ecosystems. Warmer temperatures will alter ecosystem processes and patterns of demand. The water requirements of crops for rainfed and irrigated agriculture will change, and this, in turn, will affect the water allocated to the environment. In particular, climate change will force governments to make explicit choices in the ecosystems that are to be protected when the availability of water changes in contested catchments and groundwater systems.

In some parts of the world, adaptation to climate change will require increased investments in new dams and other forms of water resources infrastructure, reoperation of existing infrastructure, and conjunctive operation of surface water and groundwater systems to buffer against the impacts of longer droughts and extreme floods. The downstream impacts of these investments will need to be assessed, both during strategic planning and during project preparation, design, and operations.

There are several reasons for the lack of political and institutional awareness about the downstream impacts arising from either infrastructure or land-use change. These include the restricted use of EFAs in project design and implementation and the limited adoption of EFAs as an integral part of environmental assessments. This situation reflects, in part, the challenge of identifying the downstream impacts; the absence of a common metric to evaluate the impacts; the diffuseness of the impacts across communities and over space; the absence of uniform methodology to delineate the downstream population affected by changes in flow; the absence of, or weak representation of, the affected parties in the decision-making process; the difficulty of expressing the respective impacts in financial and economic terms; and the lack of consensus about acceptable EFA methods.

In general, the debate about environmental flows is about multiple and evolving values of society. It is also a debate about asymmetrical power relationships between different groups: water user groups, upstream and downstream interests, urban and rural interests, public and private interests, regulators and the regulated community, developers and communities, as well as central and local interests. Consequently, any program to promote the inclusion of environmental flows into public decision making needs to be participatory, include biophysical and socioeconomic sciences, express the impacts in understandable ways (using both monetary and nonmonetary terminology), and be consistent with the principles of IWRM.

The World Bank and Environmental Flows

Interest in environmental flows within the World Bank has increased over the last 15 years (see box 1.2), mirroring the interest in and development of environmental flows globally. The Bank's 1993 water resources management policy (WRMP), based on the Dublin Principles, stipulated, "The water supply needs of rivers, wetlands, and fisheries will be considered in decisions concerning the operations of reservoirs and the allocation of water" (World Bank 1993). This explicitly identified downstream environmental water needs. The Bank's 2001 environment strategy underscored the linkage between water resources management, environmental sustainability, and poverty (World Bank 2001b). It emphasized the reliance of poor people on the productivity and environmental services of ecosystems and natural resources. It also emphasized that environmental concerns need to be moved up the decision hierarchy to policy and planning levels if environmental concerns, such as environmental flows, are to be incorporated effectively into project-level investments.[4] In the case of environmental flows, there needs to be a commitment to water policy and environment policy that recognizes water for the environment as a legitimate use of water and authorizes environmental flows with legally binding provisions and support for water allocation planning that includes water allocations for environmental purposes.

The 2003 water resource sector strategy (WRSS) was a turning point with the adoption of IWRM as a framework for water resources planning and management and its central message of reengagement "with high-reward/high-risk hydraulic infrastructure, using a more effective business model." It considered the environment as a special water-using sector as well as a central element of integrated water resources management. The new business model calls for the development of infrastructure in an environmentally and socially responsible manner. This, in turn, implies the need to take full account of both upstream and downstream environmental and social impacts and, where possible, avoid, minimize, mitigate, or offset their effects. This business model is aimed at reducing the uncertainties that are often associated with decision making in complex hydraulic infrastructure planning, design, and operations.

Following the WRSS, the Bank increased its support for environmental flows via individual infrastructure projects, in river and lake basin management and development, in sectorwide programs, and in development policy lending. The Bank–Netherlands Water Partnership Program's windows on environmental flows, river basin management, dam development, and other areas were opened in 2000 to provide support to World Bank operations on a demand-driven basis. In 2003 a series of World Bank technical notes on environmental flows were prepared to support operations (Davis and Hirji 2003a, 2003b, 2003c).

BOX 1.2
Policies, Strategies, and Resources for Supporting Integration of Environmental Flows in World Bank Operations

The following policies and strategies support the Bank principles in water and the environment:

• The 1993 WRMP stipulated, "Water supply needs of rivers, wetlands, and fisheries will be considered in decisions concerning the operations of reservoirs and the allocation of water."

• The 2001 environment strategy highlighted the environment, recognized environmental water as a legal use in water policy, and authorized its use with legally binding provisions.

• The 2003 WRSS treated the environment as a special water-using sector and central element of IWRM.

• Safeguard policies for lending operations are in place in the following areas: environmental assessment (an umbrella policy for assessing a range of impacts), natural habitats (a policy for avoiding the degradation or conversion of natural habitats unless there are no feasible alternatives and there are significant net benefits), involuntary resettlement (a policy for ensuring that resettled people are fully consulted, share in project benefits, and maintain their current standard of living), and projects on international waterways (a policy for informing affected riparian countries of proposed projects on international waterways).

Through the BNWPP environmental flows window (information available on the water Web page), support is provided by international experts to enable Bank projects to integrate environmental flow considerations into their operations. This window has supported a number of Bank operations.

The Lesotho Highlands Water Project is an example where environmental flow requirements were incorporated into the design of new infrastructure (the Mohale Dam) and the reoperation of the previous dam (Katse Dam). Projects in the Tarim basin in China and the Aral Sea in Central Asia are examples of the successful restoration of downstream ecosystems that had been severely degraded following large-scale irrigation and hydropower developments and weak water resources management. The Bank has also provided assistance in water policy reforms and river-basin-level planning in conjunction with support for infrastructure projects that have incorporated environmental flows. The shift in Bank lending from a project basis to development policy lending (DPL), programmatic lending, and sectorwide lending has provided further impetus for accelerating the mainstreaming of the environment through sector analysis as well as the use of emerging tools such as strategic environmental assessments (SEAs), country environmental assessments (CEAs), country water resources assistance strategies (CWRASs), and others.

Country water resources assistance strategies are used for defining strategic issues for Bank assistance. To date, 18 CWRASs have been produced. CWRASs from China, Tanzania, Mozambique, and the Philippines include thorough treatments of environmental flows.

Water resources and environment technical notes (available on the World Bank water Web site) provide guidance on different aspects of environmental flow science and applications:

- "Environmental Flows: Concepts and Methods" (Davis and Hirji 2003a)
- "Environmental Flows: Case Studies" (Davis and Hirji 2003b)
- "Environmental Flows: Flood Flows" (Davis and Hirji 2003c)
- "Integrating Environmental Flows into Hydropower Planning, Design, and Operations" (Nature Conservancy and Natural Heritage Institute forthcoming), prepared as part of this ESW.

An in-depth case study of the Lesotho Highlands Water Project draws detailed lessons from a complex interbasin transfer dam project between two nations (Watson forthcoming).

Hirji and Davis (2009b) review the opportunities to move environmental consideration of water resources up to the more strategic levels of policies, legislation, programs, and plans. This approach is consistent with the proposal to extend environmental flows into policies and basin plans in this ESW.

This report reviews the science of environmental flows and the global practice with environmental flows in policies, plans, and projects; it also presents a framework for better integration of environmental flows into Bank assistance. It includes 17 detailed case studies analyzed using a consistent methodology that will be published separately.

In 2007 the Bank elevated its commitment to sustainable infrastructure investments by integrating two vice presidencies working on infrastructure, the environment, and social, agriculture, and rural development within a Sustainable Development Network (SDN) to ensure a more holistic approach to development. The SDN vision not only calls for mainstreaming the environment, but also embodies environmental sustainability as a core element of the Bank's work. This commitment is reflected in the recently updated Infrastructure Action Plan approved by the Bank in 2008, the 2006 Agriculture Water Management Initiative (World Bank 2006b), and the Clean Energy Development Framework.

This report develops a framework for more systematically incorporating environmental flow considerations into Bank assistance with water policy reform, support for river basin and watershed planning and management, and investments in water resources infrastructure. It contributes to a more effective business model for reengaging in high-reward, high-risk hydraulic investments. It supports the integration of environmental flows into DPLs and water-centered sectorwide assistance and programmatic lending. It also supports the objectives of several Bank initiatives—the Infrastructure Action Plan, the Agriculture Water Management Initiative, and the climate change and water economic and sector work (ESW), as well as the Strategic Framework for Climate Change and Development—to provide environmentally sustainable investments in hydropower, water supply, agricultural water management, and flood management systems. Overall, it supports integration of the SDN vision into Bank operations.

Objectives of the Report

The overall goal of the analysis presented in this report is to *advance the understanding and integration in operational terms of environmental water allocation into integrated water resources management.* In this regard, this report complements the recently completed report on strategic environment assessment and integrated water resources management and development (Hirji and Davis 2009b).

This report has the following specific objectives:

- Document our changing understanding of environmental flows, both by water resources practitioners and environmental experts within the Bank and in borrowing countries
- Draw lessons from experience in implementing environmental flows by the Bank, other organizations with experience in this area (United Nations Development Programme, United Nations Educational, Scientific, and Cultural Organization, United Nations Environment Programme, International Union for the Conservation of Nature, International Water Management Institute, Natural Heritage Institute, the Nature Conservancy, and Worldwide Fund for Nature), and a small number of developed countries (Australia, Canada, the European Union, and the United States), and developing countries and regions (Central Asia, China, India, Lesotho, the Mekong basin, Senegal basin, South Africa, and Tanzania)
- Develop an analytical framework to support more effective integration of environmental flow considerations for informing and guiding (a) the planning, design, and operations decision making of water resources infrastructure projects; (b) the legal, policy, institutional, and capacity development related to environmental flows; and (c) restoration programs
- Provide recommendations for improvements in technical guidance to better incorporate environmental flow considerations into the preparation and implementation of lending operations.

The report is written primarily for World Bank task team leaders and water resources and environmental specialists engaged in water policy dialogue, river basin and watershed planning and management, and water resources investment planning, design, and operational decision making for investment lending. Other audiences include professional organizations, professionals from development organizations and nongovernmental organizations (NGOs), and client countries engaged in water policies, plans, and projects.

Methodology

The analysis draws on a variety of sources of information. The international literature provided information on current issues and approaches to environmental flows in both developed and developing countries. This was supplemented by information contained in several Bank documents and published articles (Davis and Hirji 2003d).

The changing perception of environmental flows within the World Bank is drawn from a review of the Bank's activities in supporting environmental flows in lending operations and technical assistance (Hirji and Panella 2003). In this report, select water-related infrastructure projects that were prepared during the 1990s were examined to see if there had been an increase in recognition of environmental flow issues following the 1993 WRMP. The Bank's CWRASs were analyzed for their recognition and integration of environmental flows. Assistance under the Bank–Netherlands Water Partnership Program (BNWPP) environmental flows window was also reviewed.

The main source of information for the lessons on implementing environmental flows came from an in-depth analysis of 17 case studies covering water policy, catchment and basin plans, and investment projects. Many of the case studies contain components that represent the world's best practice in the inclusion of environmental flows into water policy and river basin plans and in the conduct and implementation of EFAs in the development of new infrastructure and the reoperation and rehabilitation of existing infrastructure. Eight of the case studies describe projects supported by the World Bank. These case studies were analyzed with a uniform methodology both to evaluate the effectiveness of environmental flow programs and to explain factors (or institutional drivers) that may have contributed to those outcomes, as well as to identify lessons for implementing environmental flows in a variety of settings. These sources were supplemented by information drawn from other environmental flow projects supported by the Bank and other organizations, as well as technical assistance for environmental flows provided through the BNWPP[5] environmental flows window, and from a broad review of the environmental flow programs of international development organizations and NGOs. A separate stand-alone technical guidance note on integrating environmental flows into hydropower planning, design, and operations decision making was commissioned as a key input into the economic and sector analysis.

Organization of Report

The report consists of an overview; nine chapters, organized in four parts; and five appendixes. This chapter provides the context and justification for the sector analysis and outlines the methodology used in the analysis. Chapter 2 provides an introduction to environmental flows, including the reliance of downstream communities on flows; the definition of environmental flows; ecosystem services; the extent to which environmental flows are formally recognized in different countries; the incorporation of environmental flows into polices, plans, and projects; the linkage between environmental flows, IWRM, and environmental assessment at tactical and strategic levels; and the methods employed to assess environmental water needs. Chapter 3 discusses the adoption of environmental flows in the work of the World Bank. It includes a brief analysis of the evolution of acceptance of environmental flows within the Bank, the inclusion of environmental flow concerns into CWRASs, assistance under the BNWPP environmental flows window, and partnerships with other international development organizations providing assistance in environmental flows. Chapters 4 through 7 describe the analysis of the 17 environmental flow case studies. Chapter 4 describes the criteria used to analyze the case studies. Chapters 5 through 7 contain the findings from the policy-, plan-, and project-level case studies, respectively. Chapter 8 summarizes the achievements to date in integrating environmental flows into water resources decision making and the key challenges remaining. A framework for the effective integration of environmental flows into Bank operations is presented in chapter 9.

There are five appendixes. Appendix A contains the Brisbane Declaration on environmental flows. Appendix B summarizes the design options for releasing environmental flows from dams. Appendix C contains background on environmental flows. Appendix D describes the integration of environmental flows into CWRASs. Appendix E describes the contributions of the major international development organizations and NGOs providing environmental flows assistance to developing countries and their contact information.

The case studies have been published separately (Hirji and Davis 2009a).

Notes

1 Flows into marine waters are essential for many important ecosystem processes in estuaries and near-shore areas. Perceptions that these flows represent wasted water are slowly changing.

2 The Brisbane Declaration, following the 2007 River Symposium and Environmental Flows Conference in Brisbane, Australia, issues a call to action addressed to all governments.

3 Typical impacts upstream of an impoundment would be related to the conversion of a terrestrial into an aquatic habitat and could entail relocation, resettlement, or compensation of land and assets to be inundated or affected by inundation.

4 The environment strategy proposed that strategic environmental assessments be introduced as tools for this purpose.

5 The Bank–Netherlands Water Partnership Program is a programmatic trust fund established in 2000 to increase water security through the sponsorship of novel approaches in integrated water resources management and thereby contribute to the reduction of poverty. The environmental flows window has provided technical assistance to 13 countries in preparation for, or during the implementation of, lending activities.

PART II

Environmental Flows: Science, Decision Making, and Development Assistance

CHAPTER 2

Environmental Flows in Water Resources Decision Making

EQUITABLE ACCESS TO ADEQUATE QUANTITIES of good-quality water is central to growth and sustainable development. Water is a vital input to livelihoods and to most economic sectors—dryland and irrigated agriculture, livestock, forestry, inland and estuarine fisheries and aquaculture, national parks, hydropower, industrial and mineral production, transport, and tourism—in both developed and developing countries.

Access to water is also central to alleviating poverty and achieving the Millennium Development Goals (MDG). One of the MDG targets is to halve the proportion of people without sustainable access to safe drinking water and sanitation by 2015, while other MDG targets implicitly require access to clean, safe water. This requires provision of both good-quality water and sufficient quantities of water for subsistence.

The development of water resources through dams (small and large), interbasin transfers, aquifer storage and recovery, levees and dikes, and boreholes provides a buffer against climate variability. Developed countries typically have invested substantially in storage infrastructure and have much greater water storage per capita than most developing countries (see figure 2.1), even though developing countries, especially in the tropics, face much greater climate variability than developed countries. Not only do developing countries have less per capita water storage, which increases their vulnerability to extreme climate shocks such as floods and droughts, but their existing water resources infrastructure is often unreliable, poorly maintained, and

FIGURE 2.1
Water Storage per Capita in Select Countries

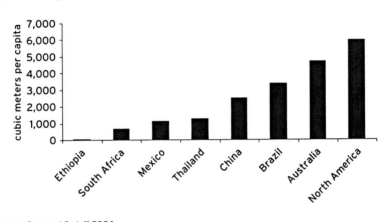

Source: Grey and Sadoff 2006.

incapable of providing buffering capacity against floods and droughts (Mogaka and others 2004). Water resources infrastructure can also, potentially, play an important role in adapting to climate change by providing water storage during extended dry periods and buffering from floods during wet periods and extreme events.

Water resources managers have been dealing with the issue of regulating flows for many years in relation to navigation, production of electricity, protection of commercial fisheries, floatation of logs, and protection against flooding. Environmental flows are increasingly being recognized as a vital part of the equation in order to maintain healthy, productive, and sustainable river and groundwater systems. South Africa's National Water Act (1998) recognizes the importance of ensuring a minimum quantity of water to sustain the ecosystem services on which many human activities depend. Other countries have similar provisions in their water policies and laws, generally mandating water for essential human and ecosystem needs. Many other countries are considering policy reforms related to environmental water.

The development of water resources has often altered the flow regimes of rivers,[1] affecting ecosystems and contributing to the decline of many species (see figure 2.2), and resulted in adverse impacts on communities downstream of the development. Abstractions of water for irrigation, water supply, or interbasin transfers for any type of use reduce the total volume of flows, while dams and other barriers also change the pattern of flows. It is important to recognize that changes in the pattern of flows—shifts in the seasonality of flows, prevention of floods reaching floodplains, maintenance of relatively high levels of flows during traditionally low-flow periods, among others—can be as disruptive to downstream ecosystems as are changes in the total volume of flows. Appendix C provides details of the current understanding of the impacts of these developments on downstream ecosystems and the development of environmental flow methodologies.

FIGURE 2.2
Changes in Freshwater Species Populations Indices, 1970-1999

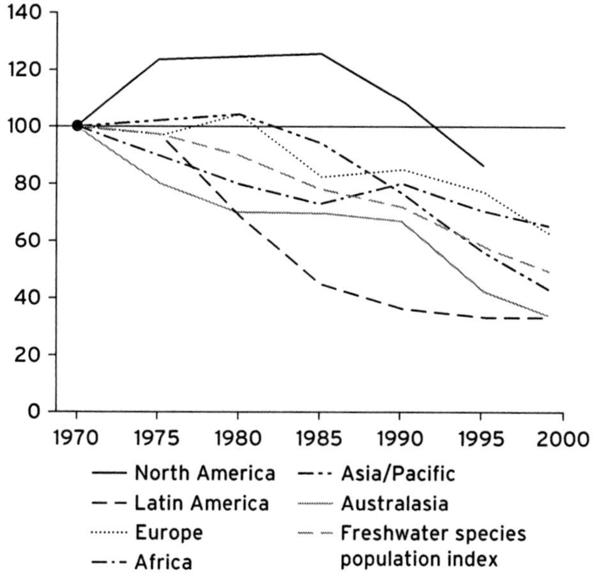

Source: World Wide Fund for Nature 2000.

While there is no international agreement specifically concerned with environmental flows, the topic is included in several other global and regional agreements. Two of note are the United Nations Convention on the Law of Non-Navigational Uses of International Watercourses—a global agreement dealing with various aspects of river utilization and management—and the Convention on the Protection and Uses of Transboundary Watercourses and International Lakes. The latter contains environmental provisions. The International Union for the Conservation of Nature (IUCN) provides a fuller treatment of international treaties and agreements dealing with water resources management and environmental flows (Scanlon, Cassar, and Nemes 2004).

The 2000 report of the World Commission on Dams provides a sharp focus on both upstream and downstream ecosystem needs and associated social impacts (World Commission on Dams 2000). It states, "Among the many factors leading to the degradation of watershed ecosystems, dams are the main physical threat, fragmenting and transforming aquatic and terrestrial ecosystems with a range of effects that vary in duration, scale, and degree of reversibility." One of the report's strategic principles is sustaining rivers and livelihoods.[2] In its sustainability guidelines, the International Hydropower Association now recognizes the need to provide for environmental flows that maintain downstream ecosystem functions (International Hydropower Association 2004). The major commercial banks providing financing for project developments have adopted a voluntary set of principles[3]—the Equator Principles—governing their social and environmental responsibilities for lending to projects of more than $10 million. Under these principles, the International Finance Corporation's categorization of project risk is adopted, and all projects that fall into categories A or B have to undergo a social and environmental assessment.

Dams are not the only infrastructure investments that can affect flows. Water that is pumped directly from water bodies or is discharged into water bodies can affect the quantity, quality, and timing of flows; levees and dikes for flood protection or other purposes disconnect floodplains and wetlands from the rivers and affect physical, chemical, and biotic processes; and excessive groundwater pumping can affect river flows where there is connectivity between the river and associated aquifers.

Land-use change can also affect downstream river systems. Converting forest to annual agricultural crops typically reduces evapotranspiration and increases

---

BOX 2.1
Examples of Flow-Dependent Ecosystem Services

Provisioning services: Tonle Sap, Cambodia. Tonle Sap is a large shallow lake in the center of the Cambodian plain (ILEC 2005). The lake is filled during the wet season from the Mekong River and, depending on the wetness of the season, can expand from about 2,500 square kilometers to up to 16,000 square kilometers. The periodic flooding carries sediment-rich water from the Mekong River to the lake, which supports a complex food web. The lake basin contains extensive wetlands and flooded forests that are critical to fish breeding.

The lake is an important source of fish for the Cambodian population, providing about 230,000 tons per year, which is more than 75 percent of Cambodia's annual inland fish catch and 60 percent of Cambodians' protein intake. It supports more than 3 million people. In addition, fish migrate from the Tonle Sap to the Mekong River and help to restock Mekong River fisheries.

Regulating services: Mississippi River, United States. The natural floodplains of the Mississippi River allow floodwaters to extend laterally and so reduce the peaks of floods for downstream communities (Belt 1975). However, the floodplains have been progressively cut off from the river by levees and banks, as the floodplain has been developed initially for agriculture and then for industrial and urban development. Whole communities are now built on the flood plain. Also, the progressive constriction of the Mississippi for navigation makes floods higher; thus navigation works degrade the protection afforded by levees.

In 1973 the levees failed under the water pressure from high flows in the Mississippi River, and large areas were suddenly inundated. Although the Mississippi crested in St. Louis 2 feet above the level of the 1844 flood, the flow in the river was actually 35 percent less than the earlier flood, illustrating the importance of the floodplain for absorbing large flows. In a study of the causes of the 1973 flood, Belt concludes that the flood was "man-made" to a great extent.

runoff. It can also reduce water abstraction from shallow aquifers (because of the shallow rooting of annual crops), leading to rising water tables and potentially higher base flows to streams. Expanding cities increase storm water runoff, reduce groundwater infiltration, and increase pollution loads, all of which can affect river flows. Conversely, afforestation can also increase evapotranspiration and reduce stream flow, while straightening of river channels can speed up runoff and increase peak magnitude of floods downstream.

Water-Dependent Ecosystem Services

Rivers provide many important ecosystem services for communities in both developed and developing countries (this section draws on Millennium Ecosystem Assessment 2005). Box 2.1 illustrates the diversity of ecosystem services provided

The subsequent floods of 1993 and the very recent floods (June 2008) on the Mississippi River and the associated widespread damage to property, farmland, industry, and cities along the river have underscored the severe consequences of encroachment into floodplains, the vulnerability of towns and communities living in the floodplains, and the inadequacies of structural measures alone to manage floods. A key lesson concerns the significant impact of changes in land use on river flows and flow patterns.

Supporting services: Nakivubo swamp, Uganda. Uganda's Nakivubo swamp has been receiving partially treated wastewater from Kampala for more than 30 years (Kansiime and Nalubega 1999). It contains dense communities of papyrus and *Miscanthidium violaceum*. These plants aid in the removal of nutrients from wastewater. In the papyrus-dominated parts of the swamp, the purification efficiencies are 67 percent for nitrogen and phosphorus and 99 percent for fecal coliform. In the *Miscantheum*-dominated parts, the removal efficiencies are lower, at 55 percent for nitrogen, 33 percent for phosphorus, and 89 percent for fecal coliform.

Cultural services: Caroni swamp, Trinidad and Tobago. The Caroni swamp has considerable biodiversity and cultural value for Trinidad and Tobago, with at least 157 birds species frequenting the swamp (Trinidad and Tobago, Ministry of Planning and Development 1999). Caroni also provides roosting and breeding habitat for a significant number of migratory waterfowl between North and South America. It was especially noteworthy as a roosting site for the scarlet ibis—the national bird of Trinidad and Tobago. But the scarlet ibis has not had a substantial nesting colony in the swamp over much of the last 30 years, presumably because of the steady salinization of the swamp. The increased salinity has been caused by several developments, including water abstractions for urban water supply, highway embankments (which have drastically altered runoff patterns), siltation, and the opening of an entrance canal for visitors.

by rivers, including (a) provisioning services (such as food, water, timber); (b) regulating services (such as flood regulation, prevention of disease, disposal of wastes); (c) supporting services (such as nutrient cycling, maintenance of river morphology); and (d) cultural services (such as aesthetic values).

Many communities, particularly in developing countries, depend on these services for protein (through fish catches), productive land for agriculture and grazing, and timber for firewood. Biodiversity underpins many of these services (Millennium Ecosystem Assessment 2005).

The ecosystem services provided by rivers are disrupted by changes in the volume, quality, and pattern of flows downstream of the development activities. Thus irrigation abstractions during the dry season can isolate river pools and prevent fish migration; river regulation because of upstream dams can reduce floodplain inundation and opportunities for recessional cropping and grazing; and changes in the volume and pattern of freshwater flows can cause silting of river reaches and loss of habitat and reduce dilution of wastewaters discharged into river systems. The IUCN's *Vision for Water and Nature* calls for "leaving water in the system to provide environmental services such as flood mitigation and water cleansing" (IUCN 2000).

Other parts of the hydrologic system, such as groundwater, also provide ecosystem services (Dyson, Bergkamp, and Scanlon 2003). Apart from being the major water resource in arid and semiarid climatic regions, groundwater also supports important groundwater-dependent ecosystems such as wetlands and swamps. Many shallow groundwater systems are connected to rivers, providing base flows in the dry season and being recharged during flood events. Groundwater systems can also be affected by upstream developments. For example, forest plantations that require large volumes of water can reduce recharge to an aquifer, thereby lowering water tables and making it more difficult to provide water for stock and domestic purposes further downstream.

Freshwater flows are also vital for estuaries and marine systems. Estuaries are complex ecosystems dependent on both freshwater and marine influences. They provide breeding habitat for fish and invertebrates, often have high recreational and scenic values, and can be economically important for transport and shipping. Changes in freshwater inflows can threaten these benefits through closure of river mouths, loss of mangroves and wetland habitats, saltwater intrusion, and reductions in nutrient and sediment inflows.

Different ecosystem functions are maintained by different components of the flow regime (see figure 2.3). The particular functions depend on the river system. Typically, low flows maintain the connectivity of pools and provide for longitudinal movement along the river; small, more frequent floods (known as freshets) can trigger spawning in some species and may remove detritus; and larger, more infrequent floods can water floodplains and provide lateral

FIGURE 2.3
Components of the Flow Regime

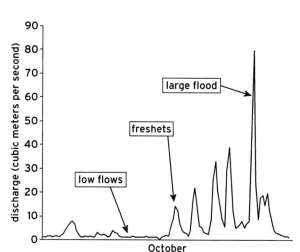

Source: Authors.

movement of sediment and nutrients to and from the floodplain. An EFA is used to identify the reliance of different ecosystems or organisms (fish, invertebrates, vegetation) on the different components of flows and their sensitivity to changes in these components. This knowledge is important when decisions are being made on allocating different parts of the flow regime to different water uses.

Environmental Flows: Adoption and Methods

Although there are various methods for undertaking EFAs, they fall into four discrete groups, namely hydrological index methods, hydraulic rating methods, habitat simulation methods, and holistic methodologies. These methods are summarized in several reviews (see, for example, Arthington and Zalucki 1998; Davis and Hirji 2003a; Dyson, Bergkamp, and Scanlon 2003; Tharme 2003).

Many early environmental flows methods were designed to protect a single species or to address a single issue. However, managing flows for a single species (and sometimes even for a single ecosystem function such as low-flow connectivity) may not result in robust aquatic ecosystems and may even fail to preserve the target species because of their dependence on a wide range of ecosystem functions (such as food webs, habitat). Consequently, holistic methodologies, which typically incorporate all components of the flow regime, are at the cutting edge of EFA methodology. Applying these methods involves a wide range of water users and sometimes includes considerations of the social and economic dependence of communities on environmental flows. Holistic methods were developed in South Africa and Australia, but are increasingly being tried in other parts of the world (see case studies 7 and 8).

The wide range of methods provides a choice of technique to suit various timetables, budgets, and purposes (see table 2.1).

Environmental Flows and Decision Making

An EFA provides the scientific basis for understanding the relationship between the different flow components and ecosystem responses. But deciding on how much and at what time(s) water should be allocated to the environment at either the river basin or project level is a decision that can only be taken in the context of all the demands on the water resource. There is no absolute quantity and timing of flows that are required for the environment or for that matter for any other use. Instead, a social choice has to be made about what uses are important, to what degree they need to be addressed, and which ecosystem services need to be preserved (and to what degree) to meet society's objectives for a particular water resource. This choice will then determine the flows that are needed to deliver those services. For example, society may decide to increase irrigated agriculture using a particular groundwater resource—at the expense of some groundwater-dependent wetlands that rely on high water tables—because the net societal benefits are greater when irrigated agriculture is increased and wetlands are decreased.

These choices have always been made in water resources planning and management. The contribution of environmental flows is that the EFA makes explicit the consequences of different choices on aquatic ecosystems and communities that depend on those ecosystem services and so leads to a more informed decision-making process. It enhances equity and sustainability in the decision-making process. This is important because downstream individuals and communities that are affected by changes in flow regimes are often relatively unorganized, powerless, and voiceless—compared to institutions and organizations that want to develop the water resource—and their traditional rights to use water are not always recognized in law. It is important to include the relationships of the communities to rivers and the needs of these downstream communities in decisions about flows.

There are two broad methods for providing environmental flows. On regulated rivers—those with water storages in their headwaters—the agreed environmental flows can be delivered through specific releases of water from the storages at the right times to mimic some of the natural patterns of flows. On all rivers—regulated and unregulated—and in all groundwater systems, controls over abstractions can also be used to retain certain components of flows. For example, cease-to-pump rules during dry periods are widely used to

TABLE 2.1
Estimated Time and Resource Requirements of Select EFA Methods

Method	Type	Data and Time Requirements	Duration of Assessment	Relative Confidence in Results	Level of Experience
Tennant	Hydrologic index	Moderate to low	2 weeks	Low	United States: extensive
Wetted perimeter	Hydraulic rating	Moderate	2–4 months	Low	United States: extensive
Expert panel	Holistic	Moderate to low	1–2 months	Medium	South Africa, Australia: extensive
Holistic	Holistic	Moderate to high	6–18 months	Medium-high	Australia, South Africa: extensive
Instream flow incremental methodology	Habitat simulation	Very high	2–5 years	High	United States, United Kingdom: extensive
DRIFT	Holistic	High to very high	1–3 years	High	Lesotho, South Africa, Tanzania: limited

Source: Davis and Hirji 2003b.

ensure that low flows are protected. A wide variety of instruments are used to provide these flows, including separate entitlements for environmental water, conditions on abstraction licenses, and dam operating rules. Where water markets have been established, environmental water can also be acquired on the market (see box 2.2).

In some cases, too many water resources are allocated to economic uses, and water needs to be recovered for the environment. This is always difficult (as illustrated in several case studies in chapter 7). There are various options for obtaining this water:

- Instituting efficiency improvements in the economic uses through technical improvements, with some of the "saved" water being used for the environment
- Reoperating infrastructure so that more efficient use is made of the water for economic purposes; this is particularly successful with hydropower dams where the operating rules have been in place for some time and may no longer fully reflect the electricity demand requirements
- Acquiring water rights from consumptive users, with or without compensation
- Augmenting water supplies through interbasin transfers, conjunctive use of groundwater, and desalinization and using some of the additional water for environmental purposes (while ensuring that there are no detrimental environmental effects from accessing the additional sources)
- Instituting demand management measures, particularly with urban water supply.

Some of these water recovery options are illustrated in the project-level case studies.

BOX 2.2
Environmental Water Trading in Australia

Under the Australian national water reforms, environmental water entitlements have the same legal standing as consumptive water entitlements and so can be traded on the market that has been established for trading consumptive water entitlements. Environmental water managers are slowly being appointed at national, state, and local levels to represent the environment in the marketplace. In principle, this tradable environmental water, termed adaptive environmental water, can be traded countercyclically. That is, it can be sold at times when the price of water is high (typically during a drought when the environment would not normally receive much water) and bought back when prices are low (when water is abundant) thus helping to make the environment self-financing. Market trading of environmental water is still to be fully implemented in Australia, and these mechanisms are yet to be tested.

Environmental Flows in Policies, Plans, and Projects

As noted, there are four entry points for introducing environmental flows into water resources planning and management decision making:
- National water policy, legislation, regulations, and institutions
- River- and lake-basin-level water allocation plans, including watershed management
- Single-purpose or multipurpose investment projects
- Restoration (rehabilitation and reoperation) projects and programs.

The last two entry points share some operational and conceptual similarities. Both involve environmental flows downstream of infrastructure and are sometimes combined in the following discussion as infrastructure investments.

Environmental flows have become identified, at least within development assistance organizations, with mitigation of the impacts of dams and other water resources infrastructure. Environmental assessment of proposals to build new dams or other infrastructure or to rehabilitate existing infrastructure should include an assessment of the potential downstream environmental and social impacts.[4]

Dealing with environmental flow issues only when development projects are proposed is unlikely to lead to equitable or efficient allocation decisions. At these times, the major decisions—for example, about siting and sizing structures (Ledec and Quintero 2003) or incorporating features such as multiple outlet valves—have often been taken, and there is usually limited flexibility to influence major decisions. Such decisions are almost always inefficient in the long run. This is illustrated by (a) the Lesotho Highlands Water Project (case study 14), where the Mohale Dam outlet valves had to be resized and a new valve had to be added to Katse Dam to accommodate higher EFA releases and (b) the Lower Kihansi Hydropower Project (case study 15), in which the process of granting and enforcing the final water right was highly contested.

More equitable development outcomes are likely to be achieved if environmental flows are considered at an earlier, more strategic stage in the decision-making process. For environmental flows, this means that environmental flow allocations should be included in river basin plans and backed up with national water policy.

There are three reasons for incorporating environmental flows into national water polices:
- Policies give legitimacy to environmental flows and thereby shift the focus of project-level discussions to the quantities and timings of water for the environment rather than on whether environmental flows are a legitimate use of water.
- Policies can be used to specify the priority to be assigned to environmental water allocations compared to other water uses.
- The procedural requirements (notification requirements, institutional responsibilities, timing, participation, and relationship to other instruments such as

environmental impact assessments and strategic environmental assessments) can be spelled out in the policy, ensuring that plan-level or project-level environmental flow studies are carried out proficiently.

The policy provisions, in turn, are often incorporated in legislation to give them force when being implemented. A water resources strategy is then typically used to spell out the steps to be followed and to identify the institutions responsible for implementing the policy and legislation. A water resources management institution, such as a river basin organization, is thereafter charged with implementing the relevant parts of the water policy and law.

Basin-level water allocation plans, drawn up under water resources legislation, identify the rights of different groups to use water resources and so should include environmental flow requirements. This means that water will be provided to maintain important ecosystem assets and functions during subsequent water management operations; this can include a formal recognition of traditional rights to water. Water allocation plans not only reduce tensions in water-scarce regions by making rights and their associated conditions explicit, but also provide a foundation for decisions on development activities that require access to water resources. They can include areas to be protected from development as well as areas that may need to be restored because of their ecological significance and importance for downstream flows. Water resources policies that contain sections on the provision of environmental water (case studies 1–5) usually require that environmental water be included in basin-level water allocation plans (Dyson, Bergkamp, and Scanlon 2003).

Thus making provisions for environmental water requirements when investment projects are being planned should be the culmination of a hierarchy of more strategic decisions about environmental water requirements rather than a one-off decision being made without considering the broader context of water management.[5]

Environmental Flows, IWRM, and Environmental Assessment

Integrated water resources management considers the environment as a legitimate use of water and integrates environmental flows into the implementation of IWRM. That is, environmental flows should not need to be promoted specifically if IWRM is properly adopted. However, most developing countries lack the resources to put IWRM into practice; the reality is that there has been only limited practical implementation of IWRM in the developing world. According to a recent World Bank analysis, environmental flows are seldom considered (Hirji and Davis 2009b). This report squarely places environmental flows in the context of IWRM so that environmental water requirements to support downstream water use becomes an integral part of both strategic (policy and basin planning) and tactical (project-level) multisectoral decision making (see figure 2.4).

FIGURE 2.4
Hierarchy of Decisions Leading to Project-level Environmental Flow Allocation

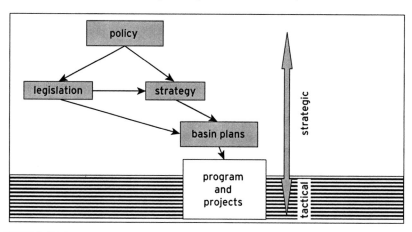

Source: Authors.

Environmental impact assessments (EIAs) are now widely accepted planning tools used to integrate environmental concerns systematically into decisions about investment projects. Most developing countries have adopted legislation requiring EIAs for infrastructure investments, and development partners (such as the World Bank and regional development banks) require an assessment of environmental and social impacts for projects they are funding. The World Bank has produced the *Environmental Assessment Sourcebook* and updates covering a wide range of topics related to environmental assessment, including those covering various water sectors, to assist with these assessments (World Bank 1991). These have been complemented by the environment and water resources technical notes, which address environmental flows, among other issues.

Undertaking an environmental impact assessment after a project has been designed is often too late in the decision-making process. The most important decisions that ultimately influence the impacts were taken much earlier, when policies, programs, and plans were being formulated and approved. Consequently, strategic environmental assessments were developed as a tool to move environmental considerations earlier in the process from project-level analysis to the levels of policies, legislation, strategies, programs, and plans.

In principle, the downstream impacts from infrastructure investments should be assessed as part of project planning and design studies, including EIAs or other appropriate planning instruments. However, in practice, these downstream impacts have not always been fully recognized or accounted for, and their assessment has often arisen as a separate process, through EFA, specifically to fill this gap. As a consequence, important opportunities for informing key design decisions are often lost. Water resources planners, EIA practitioners, and social scientists need to recognize the importance of impacts that arise downstream of projects and include EFA techniques in their toolkits, so that EFA is effectively absorbed into planning studies, EIAs, and SEAs.

Notes

1 The term "river systems" refers to a river and the hydrologic features connected to it, such as wetlands, floodplains, and estuaries.

2 The principle includes the recognition that "releasing tailor-made environmental flows can help maintain downstream ecosystems and the communities that depend on them."

3 See http://www.equator-principles.com/ for details.

4 The World Bank's 10 safeguard policies require an assessment of a range of potential environmental and social impacts.

5 See the Nature Conservancy and Natural Heritage Institute (forthcoming), which says, "Societal objectives will be best met when regional development plans, which set broad regional, national, and/or river basin objectives for water and energy development and environmental protection, are paired with more detailed local-scale environmental assessments for individual dams or cascades of dams on specific rivers."

CHAPTER 3

Environmental Flows and the World Bank

THE CONSENSUS EMERGING from the 1992 Dublin Conference on Water and the Environment and the 1992 Environmental Summit at Rio de Janeiro influenced the World Bank and development partners to increase their assistance to developing countries regarding environmental issues in general.

The World Bank, like the global water community, recognized the importance of environmental impacts that occurred downstream and upstream of water resources developments and translated this into operations using different tools and avenues. The 1993 WRMP marked the beginning of the Bank's commitment to environmentally sustainable water resources development: "more rigorous attention to minimizing resettlement, maintaining biodiversity, and protecting ecosystems in the design and implementation of water projects."

The evolving understanding of the importance of downstream environmental issues is shown by an analysis of select Bank-funded dam projects implemented before and after 1996 (given project preparation times, this year was chosen as representing the earliest date when the 1993 policy was likely to have been influential in leading to more recognition of environmental issues in preparation of dam-related projects). The main objective of the analysis was to see if more attention was given to downstream flow-related impacts after the 1993 policy was approved.

Projects that involved construction of a new dam or rehabilitation of an existing dam[1] and where the project appraisal documents (PADs), staff appraisal documents

(SARs), and EIAs were available in English were selected for analysis. At least one project was selected from each region. Twenty-eight dam projects approved before 1996 and 10 approved after 1996 were selected for analysis.

The PADs, SARs, and EIAs were examined for their inclusion of potential upstream and downstream impacts. The impacts were restricted to the biophysical impacts from changes in flow regime downstream of the dam and changes in the level of the impoundment upstream of the dam (see table 3.1). To be selected, the impacts had to be clearly identified as arising from the dam itself and not from the construction activity.

The depth of consideration of these impacts was classified as cursory (typically one or two sentences in passing), considered (at least a paragraph of specific consideration), or detailed (a detailed assessment, sometimes in quantitative terms). There is considerable subjectivity in making these assignments, but they nevertheless indicate the extent of consideration of biophysical issues in the project preparation documents.

Figure 3.1 shows the number of times an upstream or downstream issue was mentioned in the documents. The number of mentions depends on the number

TABLE 3.1
Biophysical Impacts Included in the Analysis of World Bank-Funded Dam Projects

Biophysical Impact	Upstream of Dam	Downstream of Dam	Comment
Excessive sedimentation	√	√	
Flood control	√	√	Generally beneficial downstream; includes detrimental upstream flooding
Modified fisheries	√	√	Generally beneficial upstream; generally detrimental downstream
Aquatic weeds	√		
Modified irrigation activities	√	√	
Saltwater intrusion		√	
Modified floodplain watering		√	
Bank and shoreline erosion	√	√	
Loss of biodiversity	√	√	
Modified groundwater recharge		√	
Loss of aquatic habitat	√	√	
Alteration to hydrology	√	√	References to changes in flow regime not specifically linked to a biophysical impact for downstream impacts

Source: Authors.

FIGURE 3.1
**Number of Examinations of Upstream and Downstream Issues
in Dam-related Project Documents**

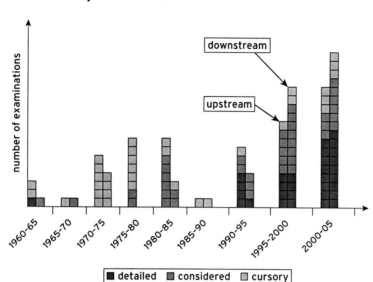

Source: Authors.
Note: No downstream issues were identified for the 1975–80 period.

of projects examined in each time period, so comparisons between time periods are neither important nor meaningful. However, the ratios between upstream and downstream issues within a period reveal the relative emphasis accorded to the regions above and below a dam.

Compared to upstream issues, downstream issues are seldom mentioned in the project preparation documents prior to 1990. Only 27 percent of the issues prior to 1996 occurred downstream of the dam; after 1996, 51 percent of issues were downstream of the dam. Over the period 1990–95 and afterward, there was also a dramatic increase in the depth of assessments. Only 2 out of 38 issues were considered in detail prior to 1990, but 30 out of 68 were classified as detailed after 1990. These results are consistent with a significant increase in the depth of treatment of environmental and social issues generally, together with a growing recognition of downstream effects of dams following implementation of the 1993 water resources management policy.

Country Water Resources Assistance Strategies

The Bank introduced country water resources assistance strategies in 2003 to bring coherence to its support for water management across the resources and service spectrum within specific countries and regions. CWRASs identify key strategic water resources issues where the Bank can play an important role in assisting with water resources development and management, including environmental issues.

The Bank has now produced 18 CWRASs. They analyze the major water resources issues facing a country or region and develop a strategic approach that countries can use to tackle these issues.

The CWRASs were analyzed to see the extent to which environmental flow issues were recognized by the countries and the Bank and whether a strategy to tackle these issues was included in the CWRAS recommendations (see table 3.2). Specifically, the analysis assessed whether the CWRAS did the following:

- Proposed inclusion of environmental flows issues in national or regional water policy and legislation
- Proposed inclusion of environmental water requirements in basin-level water plans
- Proposed measures to introduce or strengthen downstream EFAs when assessing new infrastructure projects
- Identified opportunities for providing downstream flows when rehabilitating existing dams and other infrastructure
- Proposed technical training in EFA and related procedures.

Appendix D provides a summary of these analyses.

There was no expectation that environmental flow issues should be mentioned in all CWRASs. Clearly some countries do not face water stress, and in others, environmental flow issues, even if present, may not be a priority. Nevertheless, the extent to which environmental flows are discussed in the CWRASs and incorporated into the recommended Bank strategy serves as another barometer of the degree to which this issue is being recognized and mainstreamed into water resources planning and management.

Four CWRASs (Cambodia, Dominican Republic, Peru, and Republic of Yemen) made no mention of environmental flows or equivalent concerns. The Islamic Republic of Iran CWRAS had only a passing mention through a recommended action: "Preparing and compiling necessary guidelines for studying the impacts of executing the water resources development plan on water quality and aquatic ecosystems."

In contrast, the CWRASs for some countries (such as China, the Philippines, and Tanzania) included thorough treatments of environmental flows. The China CWRAS described the overexploitation of groundwater, particularly in the Hai basin, and the overuse of surface water resulting in inadequate environmental flows in much of northern China. This, along with water pollution, is leading to the decline in and deterioration of water resources and damage to freshwater and coastal environments. The CWRAS recommended that environmental protection, including environmental flows, be one of the major themes of future Bank development assistance.

TABLE 3.2
Inclusion of Environmental Flow Issues in CWRASs

Country	Regulation by Infrastructure	Excessive Abstraction	Excessive Discharge
Bangladesh	Brief mention of effects of upstream regulation by dams	Brief mention of effects of upstream abstractions for irrigation	
China	Considerable attention to better environmental water management	Overuse of groundwater and surface water resulting in damage to freshwater and coastal environments	
East Asia and Pacific region		Proposed water allocation plans that give priority to environmental water needs	
Ethiopia	Minor mention of need for environmental flows from hydropower dams		
India		General comment that attitudes need to change to maintain instream flows for environmental benefit	
Iran, Islamic Rep. of	Statement that the development of dams will lead to deterioration of wetlands		
Iraq	Recognition that upstream dams have altered flows to the marshes, reducing flood pulses	Recognition that agricultural and urban water extractions have reduced flows	Inflows of drainage waters have affected flow regimes to the marshes
Kenya	Brief mention of downstream environmental impacts of dams	Brief mention of ecosystem degradation from water withdrawal at Lake Naivasha and on Tana River	
Mekong Region	Inclusion of the effects of proposed dams on flows	Inclusion of the effects of expansion of irrigation	

(continued)

45

TABLE 3.2
Inclusion of Environmental Flow Issues in CWRASs (continued)

Country	Regulation by Infrastructure	Excessive Abstraction	Excessive Discharge
Mozambique	Detailed consideration of effects of dams on flows		
Pakistan	Recognition that dams are reducing floods to wetlands and Indus delta	Mention of the effects of withdrawals on delta and riverine wetlands	
Philippines	Requirement that environmental flow assessments are needed for new infrastructure	Recognition that environmental flows affect surface and groundwater systems	
Tanzania	Description of problems from dams and flow-related conflicts	Description of the need to ensure that water is retained for the environment	

Source: Authors.
– Not included.
Note: The CWRASs for Cambodia, Dominican Republic, Honduras, Peru, and Republic of Yemen did not discuss environmental flow issues.

Policy and Legislation

Apart from China and Tanzania, no CWRASs mentioned the inclusion of environmental provisions into national water policies and laws. In both these cases, the countries had already established the relevant instruments. The 2002 Water Law in China includes provisions for environmental and ecological protection, and the 2002 Tanzanian national water policy requires that environmental flow provisions be made in basin-level planning.

No CWRAS identified the importance of having national policy or legislation to mandate environmental flows at basin or project levels, and consequently the CWRASs did not call for Bank support at this level. Although the Islamic Republic of Iran CWRAS did identify the need for a new water policy and governance structure, it did not specifically include environmental water requirements within the new policy.

Planning

Six of the CWRASs identified the need to build environmental flow provisions into basin-level water plans, although only the CWRASs for China and the Philippines linked environmental water requirements to planning groundwater use. The Pakistan CWRAS said that environmental flows are currently being established in the delta of the Indus. One of the proposed pillars of assistance is to promote IWRM, including provisions for environmental flows (see box 3.1). The Philippines CWRAS called explicitly for considering environmental flows in rivers and estuaries and for maintaining groundwater levels; it also called for Bank assistance in the development of basin plans that include environmental flows. The China CWRAS called for national guidelines for comprehensive river basin planning that include a requirement for considering environmental flows in rivers providing water to ecologically important areas.

Several CWRASs deal with transboundary water management. The Bangladesh, Mozambique, Tanzania, and Iraq CWRASs described efforts to promote transboundary water planning, including provisions of flows to maintain downstream ecosystems. The Iraq CWRAS described the drying of the Mesopotamian marshes, partly as a result of upstream water use by Turkey and the water developments proposed by Syria. The Mekong CWRAS recognized the need for countries in the region to better coordinate their development plans. While the Mekong CWRAS was optimistic that environmental sustainability is possible along with development, it provided few explicit proposals to achieve this.

New Infrastructure

Ten of the CWRASs discussed the need for further infrastructure development. Only four of these included an explicit recognition of the need to provide for downstream environmental water needs when planning this infrastructure, although

---

BOX 3.1

Environmental Flows to the Indus Delta

The Water Apportionment Accord was signed in 1991 by the Pakistan provinces to distribute the waters of the Indus River among the provinces and command areas. It established water rights and protected future water rights, including the effect of future storage. A formula was made available for sharing river flows. The accord also included the need for certain minimum flows (escapages) to the sea to check seawater intrusion. The provinces held different views about the necessary flows; it was decided that further studies would be undertaken to establish the minimal escapages needed below Kotri Barrage, the main regulating structure on the lower Indus River.

Three studies were agreed on after intense negotiation among the provinces. The first study determined the minimum flow below Kotri Barrage to control seawater intrusion into the delta. The second addressed environmental impacts from river water and sediment flows and their seasonal distribution below Kotri Barrage. The third addressed environmental concerns about a wide range of issues related to the management of water resources upstream of Kotri Barrage. These reports were then assessed by an independent panel of experts.

The panel of experts recommended an escapage at Kotri Barrage of 5,000 cubic feet per second throughout the year to check seawater intrusion, accommodate the needs for fisheries and environmental sustainability, and maintain the river channel. They recommended that 25 million acre-feet in any five-year period be released in a concentrated flood flow to maintain sediment supply to the mangroves and coastal zone. They also recommended that any further upstream development of storage preceded by a full EFA.

Source: González, Basson, and Schultz 2005.

---

some (such as the Philippines CWRAS) emphasized the importance of environmental flows elsewhere in the document, implying that these issues would be included in an environmental assessment of new infrastructure. In contrast, the China CWRAS placed a high priority on strengthening environmental management, including environmental flow provisions. It saw loans for physical infrastructure as a vehicle for integrating the particular project with basinwide management in order to address the full range of water issues including environmental objectives.

Infrastructure Rehabilitation

Five CWRASs foresaw the need to rehabilitate existing water resources infrastructure, especially dams that had been poorly maintained. The Pakistan CWRAS identified the need to rehabilitate a large stock of old infrastructure as well as the need to build new dams. Much of this old infrastructure has resulted in downstream environmental problems; for example, the India CWRAS described the "environmental debt" that hangs over the country's water infrastructure and is still not widely recognized by senior decision makers. Only two CWRASs specifically stated that there are opportunities to remedy these issues. With the renegotiated ownership of the Cahora Bassa Dam in Mozambique (together with two new project proposals), for example, there is an opportunity to restore environmental flows to rehabilitate some of the damaged downstream ecosystems on the Zambezi River.

The Bangladesh CWRAS called for environmental flow studies on the Ganges River to assess the human and environmental impacts of existing and proposed infrastructure and to provide a scientific basis for determining environmental flows. However, since this infrastructure was in India, the CWRAS did not contain any rehabilitation proposals. The Kenya CWRAS recognized that existing dams have caused downstream social and environmental problems and called for better project planning in the future, but it did not envisage any rehabilitation or changes in operating rules.

Training

Few of the CWRASs mentioned the need to develop skills and experience in environmental flows. The Tanzanian CWRAS is distinctive in its inclusion of a comprehensive environmental flows training program (see box 3.2). The Pakistan CWRAS also included the training of a new generation of water resources specialists in all aspects of water planning and management, presumably including environmental flow provision, while the Iran CWRAS called for training on a range of topics, including "water and the environment."

Bank-Netherlands Water Partnership Program

The government of the Netherlands established the Bank–Netherlands Water Partnership Program following the Second World Water Forum in the Hague in March 2000. The BNWPP supports World Bank operations and promotes innovative approaches for integrated water resources management in the Bank's client countries and the broader development community. The BNWPP operates through subprograms (called windows) corresponding to various IWRM topics. Each window has a team of experts who provide assistance to World Bank task managers to improve the quality of their ongoing operations. The BNWPP also supports activities that generate knowledge and development of best-practice materials (World Bank 2001a).

One of the windows specifically deals with environmental flow allocation.[2] Its objective is to assist World Bank client countries to integrate environmental

---

BOX 3.2
Proposed Environmental Flows Program for Tanzania

The 10-point plan (supported under the BNWPP environmental window) includes a wide range of activities required to build a long-term sustainable program in Tanzania that supports institutionalization of EFA into water resources planning and management decision making. Some activities are large and will take several years; others are small and can be implemented rapidly. These activities are very broadly in the chronological order in which they would be undertaken, although some may overlap or be done in parallel:

1. Training course: "getting experience of environmental flow frameworks and methods"
2. Defining an assessment framework: "turning policy into action"
3. Trial application of environmental flow methods: "practicing what we've learned"
4. Visits to foreign case studies: "seeing what others have done"
5. Technical workshop or symposium: "discussing our techniques"
6. Technical support: "checking what's been done"
7. National database: "assembling a library of knowledge"
8. Networking: "sharing experience"
9. Research: "improving our understanding"
10. Communications strategy: "spreading the word."

Source: World Bank 2006c.

---

flow considerations into water resources management and project development activities. The window draws from a panel of international experts who are available to assist Bank staff and borrowing country staff with environmental flow issues. Not all activities draw on these experts. The window has provided support to a number of Bank projects where environmental flow requirements need to be assessed and incorporated into decision making. Many of these activities are in their early stages of implementation; table 3.3 summarizes some of the main assistance activities.

Overall, the BNWPP environmental flows window has provided good support to Bank activities. Support to the Lesotho Highlands Water Project resulted in an analysis of the economic impacts of different environmental flow options as well as a thorough review of the lessons emerging from this project. Support for the Ningbo Water and Environment Project in China led to an assessment of the downstream impacts of the development. Support to the Lower Kihansi Environmental Management Project helped to develop longer-term capacity in Tanzania in environmental flows. The expert funded to introduce environmental flows to the Mekong River Commission was subsequently hired by the commission to advise

TABLE 3.3
Select BNWPP Assistance to World Bank Projects

Project	Region	Sector	Type of BNWPP Activity	Issues	Outcomes	Status
Lesotho Highlands Water Project	Africa	Water supply	Economic analysis of EFA and dissemination of environmental flow policy development and implementation results	• Assessment of environmental flow requirements • Establishment of flows following earlier treaty agreement • Redesign of outlets to accommodate required flows • Enforcement of flow agreement	• Acceptance by Lesotho of necessary environmental flows • Development of comprehensive assessment method (downstream response to improved flow transformation, DRIFT) • Successful redesign of outlet structure • Performance audit of instream flow requirement policy every five years	Completed
Lower Kihansi Environmental Management Project	Africa	Hydropower	Technical assistance in national EFA capacity-building effort	• Discovery of rare and threatened ecosystem after dam commenced • High national dependence of water for electricity production • High cost of retrofitting dam for additional flows	• Offsite breeding of threatened toad species • Agreement on environmental flow provisions and water rights for hydropower • Development of a catchment management plan • Establishment of professional training course in EFA	Phase 1 completed; Phase 2 support ongoing

(continued)

TABLE 3.3
Select BNWPP Assistance to World Bank Projects (*continued*)

Project	Region	Sector	Type of BNWPP Activity	Issues	Outcomes	Status
Mekong River Water Utilization Project	East Asia and Pacific	Hydropower, irrigation	Initial technical assistance on EFAs for the Mekong River to help the Mekong River Commission to establish and implement flow rules	• Proposed development of upstream dams in transboundary setting • High dependence of downstream fishing and agricultural production on flow regime • Divergence of development objectives among basin countries • Maintenance of significant aquatic ecosystems	• Development of hydrologic and hydraulic models • Examination of flow implications of development scenarios • Hiring of BNWPP expert as an EFA adviser to the Mekong River Commission	BNWPP support completed; Mekong EFA ongoing
Orissa Water Resources Project, India	South Asia	Irrigation	Technical assistance for establishing environmentally sensitive operational rules for Naraj Barrage	• Changes in flows to Chilika lagoon and increased sediment loads from the catchment leading to reduced exchanges with the ocean	• Development of hydrological models • Completion of technical training of Orissa State water resources staff • Improved understanding at management level of relevance of environmental flows and procedures for establishing flow requirements	Completed

Project	Region	Sector	Technical assistance	Issues	Outcomes	Status
				• Reduction in fish catches, effects on lake's biodiversity, decreased salinity, increased flooding, and weed infestation • Loss of livelihoods leading to civil disturbance		
Ningbo Water and Environment Project	East Asia and Pacific	Water supply	Technical assistance to advise on monitoring needs and environmental water requirements	• Diversion of water for urban supply leading to desiccation of the Ningbo River downstream of dam • Aesthetic and environmental issues from water loss	• Awareness of environmental flow concepts and techniques • Training in EFA methods • Development of an environmental flow scientific program	Ongoing
Mexico	Latin America and Caribbean	Water policy	Technical assistance reviewing impacts of water policies and programs on environmental needs; use of economic instruments for providing environmental water	• Ongoing reform of water policy to deal with conflicting demands • No experience with environmental water considerations	Unknown	Ongoing

(continued)

TABLE 3.3
Select BNWPP Assistance to World Bank Projects (*continued*)

Project	Region	Sector	Type of BNWPP Activity	Issues	Outcomes	Status
Rehabilitation of the Kura-Araz basin, Azerbaijian	Europe and Central Asia	Irrigation	Technical assistance on incorporating environ- mental water require- ments into new and rehabilitated infra- structure projects	• Severe degradation of wetlands and lakes	• Training sessions for managers on environ- mental flow concepts and best practice	Completed
Hydropower Umbrella Project, Ecuador	Latin America and Caribbean	Hydropower	Preliminary assessment of environ- mental flow needs for two proposed hydropower plants	• Degradation of rivers, presence of rare endangered fish species • No environmental flow policy or experience in Ecuador	• Preliminary assessment of environmental flow needs; full analysis not undertaken	Completed

Source: Authors.

on the introduction of EFA in the basin. The window has also supported various operations (hydropower, irrigation, river basin management) to introduce environmental flows and implement environmental flow assessments in other countries, including Azerbaijan, Ecuador, the Ukraine, and Uzbekistan.

There have been other window-supported activities where the results have yet to be realized. The extensive training and guidance provided to the Water Resources Department and the Chilika Development Authority in the Chilika basin (India), for example, has yet to result in the integration of environmental flow allocations in the operating rules for the Naraj Barrage, and it is not clear whether the support for water resources policy reform in Mexico will lead to the inclusion of environmental flows in the country's national water policy.

World Bank Safeguard Policies

The World Bank's 10 safeguard policies support the integration of environmental and social concerns into project design and the decision-making processes of borrowing countries (see box 3.3). These policies apply to investment lending operations, including sector investment loans, financial intermediary lending, rapid response, Global Environment Facility (GEF), and carbon finance operations. Although not covered by the safeguard policies, development policy lending operations, which are covered by operational policy (OP) 8.60/Bank procedure (BP) 8.60, are also subject to environmental and social review through the use of SEAs, poverty and social impact analysis, and other instruments. Investment lending and DPL operations are subject to the Bank's disclosure policy and provide for public consultation and disclosure. All of these policies may deal with topics relevant to downstream impacts, depending on the scope and nature of a proposed development program or project.

Extensive support material is available to help apply the environmental assessment safeguard policy (OP/BP4.01), especially the *Environmental Assessment Handbook* (World Bank 1991) and a series of updates that deal with specialized topics. In addition, the water resources and environmental technical notes on environmental flows provide a general introduction to environmental flows.

Partner Agency Collaboration

The World Bank is also collaborating with several international development organizations, including the Danish International Development Agency (DANIDA), International Water Management Institute (IWMI), United Nations Environment Programme (UNEP), United Nations Development Programme (UNDP), United Nations Educational, Scientific, and Cultural Organization (UNESCO), and the U.S. Agency for International Development (USAID). It is also working with international conservation NGOs, including the IUCN, Natural Heritage Institute

---

BOX 3.3
World Bank Safeguard Policies

OP/BP 4.01 Environmental assessment. An umbrella policy requiring environmental assessments to cover a broad range of potential impacts.

OP/BP 4.04 Natural habitats. Avoid the degradation or conversion of natural habitats unless there are no feasible alternatives and significant net benefits.

OP 4.09 Pest management. Promote environmental and biological pest management for both public health and agricultural projects. Chemical methods can be supported where justified.

OP/BP 4.12 Involuntary resettlement. Avoid resettlement where possible and, where not, ensure that resettled people are fully consulted, share in project benefits, and maintain their standard of living.

OD 4.20 Indigenous peoples. Ensure that indigenous peoples are involved in fully informed discussions so that they do not suffer adverse effects and that they receive culturally compatible social and economic benefits from Bank-financed projects.

OP 4.36 Forestry. Harness the potential of forests to reduce poverty, while integrating forests into sustainable economic development and protecting the environmental services and values of forests.

OP/BP 4.37 Dam safety. Ensure that new dams are constructed and operated to internationally accepted standards of safety and existing dams used in a project undergo safety inspection and any necessary upgrades.

OP/BP 4.11 Physical cultural resources. Avoid or mitigate adverse impacts on physical cultural resources such as valuable historical and scientific information, assets for economic and social development, and integral parts of a people's cultural identity and practices.

OP/BP 7.50 Projects on international waterways. Inform affected riparian countries of proposed projects on international waterways and, if there are objections, refer the proposal to independent experts.

OP/BP 7.60 Projects in disputed areas. Ascertain that the governments concerned agree that the undertaking of a project in disputed areas does not damage claims made by other governments.

(NHI), the Nature Conservancy, and the World Wide Fund for Nature (WWF) as well as with research organizations that offer financial or technical assistance to developing countries to undertake and implement EFAs and protect downstream ecosystems. Such assistance includes practical, longer-term technical assistance with EFAs for specific restoration work and new infrastructure projects, technical assistance and financial assistance with the inclusion of downstream flow concerns into river basin plans, shorter-term training and capacity building, and provision of resources for water resource and environmental specialists.

The Bank has collaborated with these international development organizations and conservation NGOs at several levels—global, regional, national, and basin—and has drawn on their experience and expertise in EFA and also their presence on the ground. Thus, for example, the Nature Conservancy and NHI have produced a technical note for the Bank on integrating environmental flows into the planning, design, and operations of hydropower dams (see box 3.4) as a contribution to the ESW on which this report is based. NHI also is collaborating with the GEF, the African Development Bank, and the World Bank to examine the feasibility of reoperating existing dams in order to improve their environmental performance. Appendix E highlights selective but relevant environmental flow–related work of the various organizations. It is intended to inform

---

BOX 3.4

Designing Hydropower Dams to Include Environmental Flows

Several structural and operational considerations in the development of hydropower (and other) dams can facilitate the integration of environmental flow objectives, including the following:

- Variable outlet and turbine-generator capacities
- Multilevel, selective-withdrawal outlet structures
- Reregulation of reservoirs
- Power grid interconnection
- Coordinated operations of cascades of dams
- Flood management in floodplains
- Sediment bypass structures and sediment sluice gates
- Fish passage structures.

These should be considered from the earliest stages of planning and designing the dam.

The operating objectives for dam projects are likely to change over time, in response to changing social priorities, scientific and technological advancements, and climate change. This places a premium on maintaining flexibility to modify dam operations. Many recent experiences suggest that it is possible to improve the environmental performance of existing dams (called "reoperation") in a cost-effective manner, sometimes with little or no social or economic disruption. Reoperation can be accomplished by implementing various water or energy management techniques that increase the flexibility of reservoir storage and releases such that environmental flows can be released into the downstream channel and floodplain. However, it is far easier and more cost-effective to integrate environmental flow considerations into the planning and design of dams than to modify or retrofit the design and operation of existing schemes.

Source: The Nature Conservancy and Natural Heritage Institute, forthcoming.

Bank staff of the types of activities and potential opportunities for future collaboration. However, there are opportunities to increase the level of collaboration, combining their experience in conducting EFAs and training with the Bank's experience in implementing infrastructure projects and water policy reforms.

Notes

1 Projects were excluded that involved installation of turbines and other equipment in existing dams or where the rehabilitation did not involve substantial construction that had the potential to affect flows.

2 Two other windows—river basin management and dam development—potentially would integrate environmental flow considerations. The activities supported under these two windows have not been reviewed.

PART III

Case Studies of Environmental Flow Implementation

C H A P T E R   4

Case Study Assessment

FIVE POLICY-LEVEL, four catchment- or basin-plan-level, and eight project-level case studies were selected for analysis to identify the factors that promoted or impeded successful EFA in policies, plans, and projects. The case studies were also analyzed for the drivers that initiated the EFAs and supported their implementation.

These case studies provided a diversity of institutional settings, geographic regions, and levels of economic development (Hirji and Davis 2009a). In order to draw on the best available experience, the case studies included eight cases that were supported by the World Bank and nine other cases that were not supported by the Bank but that contained components that represent international best practice.

Good-Practice Criteria

The International Association for Impact Assessment has published a set of good-practice principles for producing SEAs (IAIA 2002). These have been adapted and used to analyze 10 SEA case studies in previous World Bank economic and sector work (Hirji and Davis 2009a). EFAs are a special type of environmental assessment. Project-level EFAs are a type of EIA, and basin- or catchment-plan EFAs are a type of SEA. Consequently, the criteria used to assess the project-level and plan-level case studies were developed from these SEA criteria (Hirji and Davis 2009b). The good-practice policy-level criteria were developed around the need for policy support for implementing these plan- and project-level EFAs.

The following good-practice assessment criteria were used for the policy case studies analyzed here:

- *Recognition.* Environmental allocations were recognized in the policy (and legislation) as legitimate uses of water and necessary for the provision of ecosystem services.
- *Comprehensiveness.* All components of the water cycle were included in the policy provisions, and national and transboundary environmental flow concerns were included.
- *Environmental water mechanism.* The policy and legislation identified a mechanism for establishing environmental objectives and providing water for the environment.
- *Participation.* The policy and legislation included provisions for encouraging stakeholder participation in formulating environmental flow requirements and participating in the making and implementation of decisions.
- *Assessment method and data.* The policy and legislation provided guidance on the use of information.
- *Reviewing, monitoring, and enforcement.* The policy and legislation included provisions for reviewing, monitoring, and reporting environmental outcomes.

Both basin or catchment plans and water resources projects were assessed for the extent to which they incorporated the following:

- *Recognition.* The legitimacy of environmental flows was recognized by all parties when the plan was being formulated and implemented and the project proposal was being assessed.
- *Comprehensiveness.* All relevant components of the water cycle were included in the EFA.
- *Participation.* Stakeholders with an interest in environmental flow outcomes were engaged in the process.
- *Assessment method and data.* Recognized methods and reliable data were used in the EFA.
- *Integration.* There was integration between environmental impacts and the consequent social and economic impacts of water allocation decisions.
- *Cost-effectiveness.* The EFA methods were cost-effective, and the provision of environmental flows within the plan resulted in cost-effective outcomes.
- *Beneficial influence.* The EFA had a beneficial influence on the allocation of water for environmental purposes within the plan, as well as more widely.

Institutional Drivers

To be effective, an EFA also needs to be embedded in an appropriate enabling environment and championed and driven by powerful influences. This analysis also assessed the drivers that led to the initiation of the inclusion of environmental flows in water resources policy (policy-level case studies) and the drivers that brought about the EFAs at the basin- or catchment-level plan and project levels.

The basin- or catchment- and project-level case studies were analyzed against a set of institutional drivers that were originally identified for project-level EIAs, but that also are relevant for project- and plan-level EFAs (Ortolano, Jenkins, and Abracosa 1987). They are described in box 4.1.

BOX 4.1
Drivers for Environmental Flows in Plans and Projects

Judicial drivers. The courts have a formal role in ensuring that government organizations implement EFA provisions in the relevant legislation. Judicial drivers are used widely in the United States, where the judiciary has a constitutionally sanctioned role in reviewing government procedures.

Procedural drivers. Legislation, regulations, and guidelines provide formal drivers over the procedures to be followed when EFAs are conducted for basin water allocation plans or project impact assessments. However, procedural drivers are seldom effective without the availability of other drivers such as evaluative or professional drivers. By themselves, they can lead to well-written EFAs that are ignored. These drivers also include external agreements such as international conventions and regional agreements.

Evaluative drivers. Evaluative drivers exist when there is an institution responsible for assessing the quality of implementation of policy require-ments or plan- or project-level EFAs. These independent assessors may have the power to return catchment or basin plans or EFAs for revision, may be able to impose fines for lack of compliance with policy requirements, or may rely on publicity to generate effective implementation of policy.

Instrumental drivers. The requirements of international development partners provide an additional driver for EFAs. Thus many development partners have formal requirements for EFAs as part of the due diligence attached to loans. There can also be informal instrumental drivers operating where the development partner advocates environmental flow considerations when supporting water policy reforms. Instrumental drivers can play a central role in developing countries, where legislative and evaluative drivers are absent.

Professional drivers. The considered judgment of planners, professional associations, and other professionals undertaking policy development, catchment or basin plans, or project developments can act as a powerful driver for EFAs. Professionals can be influenced by international develop-ments in EFAs or, more broadly, in environmental sustainability.

Public drivers. These drivers rely on informed public citizens, community-based organizations, and nongovernmental organizations that are motivated and confident enough to make their views about environmental equity known to government. They may be more relevant in developed countries, which have a tradition of active public engagement in the decisions of government, but may also be important in developing countries. A stimulus is often provided by local, national, or international NGOs, who make an assessment and then inform the public.

Source: Modified from Ortolano, Jenkins, and Abracosa 1987.

However, the drivers that lead to the inclusion of environmental flows into water resources policies differ from the ones that operate for plans and projects. Environmental flow provisions are included in policy when the policies themselves are being revised, and so the policy drivers need to include both those that lead to the policy reforms as well as those that operate specifically to include environmental flows into the new policies. Three types of policy reform drivers were apparent in the case studies, and four drivers operated for environmental flow provisions (see box 4.2).

BOX 4.2
Drivers for Environmental Flows in Policies

Drivers are divided into those that encourage policy reform and those that encourage the inclusion of environmental flow considerations in the water resources policy.

In the area of policy reform,

Convening. In a federal system, the superior government can use its influence to convene and lead policy reforms even when the responsibility for the policy lies at a subsidiary level. This convening power is sometimes supplemented with financial assistance from the federal level to help the subsidiary levels of government implement the policy reforms.

Singularity. A singular event, such as a drought, can precipitate policy reforms if it is clear that the current water policy is inadequate to handle the event. While such events act as triggers for reform, there is often a backlog of issues, including provision of water for the environment, that need to be incorporated into the new policy beyond the particular deficiency that triggered the reform.

Public. Public pressure, because of perceived deficiencies in water resources management, can act as a powerful stimulus for reform.

In the area of environmental flow inclusion,

Institutional. Water managers and other professionals within government can support the inclusion of environmental flow provisions in policy because they are aware of the benefits that these flows confer on downstream environments and communities.

Evaluative. A specific organization can be identified in the policy with the oversight of environmental flow provisions to ensure that they are implemented. The organization is typically at least partially independent of government, since it is overseeing the performance of government organizations. This driver acts to implement the environmental flow provisions rather than to introduce them into policy.

Public. Where the public is concerned about the decline in downstream environments because of water abstractions and other developments, they

---

(continued)

can exert considerable pressure for environmental flow provisions to be included in policy reforms.

Scientific professional. Scientific organizations and individual scientists can use their standing in government and in the community to highlight the issues arising from disruptions to downstream flows and to propose policy provisions to help to restore downstream environments.

International developments. The proclamations from major international conventions, such as the 1992 Rio Summit, can exert considerable influence on the contents of new policies. This can manifest itself in decisions to bring in international expertise in environmental flows into countries so that they can demonstrate that they are using international best practice.

CHAPTER 5

Policy Case Studies: Lessons

AUSTRALIA, THE EUROPEAN UNION (EU), Florida (United States), and South Africa were selected for the policy case studies because they represent the major countries where environmental flows have been introduced through water policies (see table 5.1). These policies have now been implemented for several years, so they offer good opportunities for learning. Australia and the EU also provide opportunities to learn from environmental flow provisions in transboundary policy settings. The fifth case study, Tanzania, provides an example where environmental flows are required in the national water policy of a developing country. These case studies are described in Hirji and Davis (2009a).

Assessment of Effectiveness

As detailed in chapter 4, the analysis assesses projects according to good-practice criteria in the following areas: recognition, comprehensiveness, environmental water mechanism, participation, assessment method and data, and reviewing, monitoring, and enforcement.

Recognition

Assigning priorities to environmental water is an important indicator of the importance to be attached to environmental allocations. All five water policies recognize

TABLE 5.1
Characteristics of Select National Water Policies

Case Study	Country or Region	Gross Domestic Product per Capita (US$)[a]	Institutional Setting	Sector	Date Completed
National Water Initiatives	Australia	$35,990	Federation of states	Multisectoral	1994; revised 2004
Water Framework Directive	European Union	$4,089–$89,571	Union of countries	Multisectoral	2000
Florida water policy	United States	$44,970	State government within federal system	Multisectoral	1972; subsequent amendments
National water policy	South Africa	$5,390	Unitary government	Multisectoral	1997
National water policy	Tanzania	$350	Unitary government	Multisectoral	2002

Source: Authors.
a. From World Bank Doing Business 2008 site. http://www.doingbusiness.org/ExploreEconomies/EconomyCharacteristics.aspx.

the importance of ensuring that water is allocated to the environment, although the EU Water Framework Directive (WFD) treats environmental flows as a secondary issue compared to water quality issues and maintenance of ecosystem health. The priority accorded to environmental flows compared to other uses of water differs among the countries. The South African and Tanzanian policies assign explicit priority positions (first or second priority) to environmental water allocations. The Australia, EU, and Florida policies do not mandate a priority, although the Florida implementation rule implies that environmental water allocations have a high priority, except in times of drought. Although the concept of assigning the environment first or second priority in the water allocation process makes its importance clear, specific priorities in water allocation are difficult to put into practice. When water allocation decisions are made in a basin plan, tradeoffs between environmental and other water uses are inevitable. Unless there is an explicit procedure or mechanism for putting these priorities in place, it is not clear how the concept of allocation priority is to be interpreted when making these tradeoffs.

The links among environmental health, provision of ecosystem services, and human benefits need to be made explicit in the national water policies of developing countries. These links are clearly stated in the South African, Australian, and Tanzanian policies. Because the EU WFD focuses specifically on ecosystem health rather than on environmental flows, it does not make these links clear and does not link the protection of ecosystem health to human benefit. That is, the purpose of the policy is to establish a minimum level of ecosystem health, and if this requires modification of the flow regime, then environmental flows will be implemented. Similarly, the Florida implementation rule promotes minimum flows and levels to protect environmental values but does not link these to the ecosystem benefits that humans enjoy. And the minimum flows and levels are only instituted for ecosystem protection for surface waters; maintaining minimum levels for groundwater is for the physical sustainability of the resource rather than for the benefit of any groundwater-dependent ecosystem.

It takes considerable political will and administrative drive to implement environmental flow provisions. While the policies, except for the European Union WFD, all provide clear recognition of environmental flows, the implementation of these policy provisions differs considerably across the countries. The South African policy has been implemented slowly because of the extensive consultation requirements in the legislation and the reluctance of water users with existing entitlements to accept that the water resources are overallocated and that the environment is a legitimate and priority water user. In Tanzania, where the water reforms are not bound up with a major social change agenda, implementation of the water policy is moving ahead even before the water resources legislation has been passed. In Australia, even though many components of the water reforms have now been implemented, the rollout of water plans has been slower than intended (although about 120 catchment and groundwater plans have now been completed). In Florida the minimum flows and levels were not implemented until 20 years after the act had been passed (although 237 minimum flows and levels have now been established and 114 are pending).

Giving environmental water entitlements at least equal standing in law to consumptive water entitlements provides security to environmental water allocations. In South Africa, the ecological reserve, along with the basic human needs reserve, is enshrined in the law as the only water right. All other uses need permits for use after the reserve has been established. The Australian National Water Initiative (NWI) does not give an explicit priority to environmental water provisions, but it does require that environmental water be given the same statutory recognition as consumptive water entitlements. This not only places environmental water allocations on the same footing as other water uses, but also opens the door for trading in environmental water entitlements and allocations.

Water and environment policy and legislation provide legitimacy and guidance to EFAs at the plan and project levels. All policies, except the Tanzanian policy, are now supported with legislation. This provides legitimacy to the environmental water provisions of the policies and guidance about the mechanisms to use for implementing these provisions. While legislation is important, Tanzania provides an example where it is not necessary to wait for the legislation before preparing for water allocation planning. A trial environmental flow assessment has commenced in one basin, and several EFAs are either under way or being planned in other basins.

Passing a policy with provisions for environmental flows does not mean that sectoral harmonization will follow. It is essential to have the concepts of environmental flows recognized as legitimate by the professional staff of the relevant water resources and environment organizations and water-dependent sectoral institutions. The case studies illustrate the diversity of acceptance of environmental flows across different organizations. In Australia, the EU, and Florida, there is now a general acceptance of the importance of providing water for the environment. In Tanzania, the Water Resources Department under the Ministry of Water and Irrigation exhibits the greatest understanding and has taken the lead in implementing environmental flows; the environment organizations contribute but do not take the lead, although they are now working on building capacity for EFAs. Some Tanzanian ministries—typically those involved in hydropower development—and the Irrigation Department have yet to understand the importance of environmental flow considerations. Within the South African Department of Water Affairs and Forestry, a similar tension exists between the staff who advocate the importance of providing flows that maintain ecosystem services and the more development-oriented sections of the department. The same difficulty was encountered in the assessment of environmental flows for Chilika lagoon, India (case study 13), where the state Water Resources Department engineers found difficulty in grasping the ecological and social concepts behind environmental flows.

Comprehensiveness

Environmental provisions need to be comprehensive across the water cycle to include surface water and groundwater, estuaries, and near-shore regions. The five policies are comprehensive in recognizing the importance of environmental functions of the surface freshwater cycle continuum (lakes, rivers, wetlands, floodplains), although only the Australian and South African policies include the importance of controlling land uses that intercept overland and subsurface flows because of their potential to remove water that is needed to support downstream ecosystem functions. These interception activities can remove significant quantities of water and should be included in the water policies of countries where forestry or other high-water-demand land uses are prevalent.

The Florida policy gives equal weight to surface water and groundwater (in recognition of this, it uses the term "environmental flows and levels"), while Tanzanian and South African policies have sections on groundwater. However, groundwater is not explicitly included in the South African definition of the ecological reserve (van Wyk and others 2006). The 1994 Council of Australian Governments (COAG) water policy was very weak in its recognition of the role that groundwater plays in sustaining environmental functioning; this omission was recognized to cause significant consequences and was remedied in the 2004 NWI agreement. More recently, the Australian Water Act 2007 requires that the use of water in the Murray-Darling basin be subject to an integrated surface water and groundwater cap (see box 5.1).

The need for freshwater inflows to maintain estuarine ecosystem functions is absent from the Australian NWI policy and only mentioned in passing in the South African policy and legislation. The European Union WFD explicitly recognizes the importance of providing river flows into estuaries. In the Florida policy, the minimum flows and levels apply to coastal waters and estuaries. The Tanzanian policy states explicitly that water for the environment is required to maintain "the health and viability of riverine and estuary eco-systems." However, the extent to which environmental flows for estuaries are actually incorporated into water allocation plans is difficult to determine. They have been included, although not always well integrated, in some Australian catchment plans. South Africa, which does not give strong recognition to estuarine water needs in its policy, is an international leader in determining the flows needed to maintain estuarine values.

Climate change should be accounted for when establishing environmental flow provisions. None of the policies explicitly links climate change in the assessment

---

BOX 5.1
Managing the Whole Water Cycle

The consequence of focusing on surface water and neglecting groundwater is shown by the perverse outcome from the cap on surface water use in Australia's Murray-Darling basin. During the 1980s and 1990s, water abstractions grew rapidly in the Murray-Darling basin primarily to service the growth of irrigated crops. Because of concerns about the damage being done to the aquatic environment, in 1995 a cap was placed on further abstractions from surface waters beyond the abstractions that were diverted in the 1993–94 year.

While the cap has (with a few exceptions) been adhered to and surface water abstractions have remained steady at about 11,200 gigaliters a year, there has been a dramatic increase in groundwater use within the basin. Groundwater licenses have been issued that could allow the extraction of 3,261 gigaliters a year, around 34 percent of the surface water allocation. An estimated 186 gigaliters a year of stream flow have already been captured due to the growth in groundwater extraction from the introduction of the cap until 1999/2000 because of the connectivity between surface water and groundwater. This figure will grow as abstractions from less directly connected groundwater systems start to have an impact on rivers.

A review in 2000 recommended that the surface water cap be replaced with an integrated surface water and groundwater cap that was based on the water needed for ecosystem functioning, rather than water abstraction in an arbitrary year. This recommendation has now been enacted in the 2007 Water Act.

Source: Murray-Darling Basin Commission 2000.

---

of environmental flows. Climate change is included as one of the inputs to water allocation planning in the Australian NWI, but it is not linked specifically to the potential effects on ecosystem functions. In the South African policy, there is an allusion to "human activities [that] are beginning to have a noticeable impact on our climate," although this is not linked to ecosystem functioning. Climate change is not included in the European Union WFD, the Florida implementation rule, or the Tanzanian water policy. These are important omissions, since climate change is predicted to have significant effects on the availability and use of water in many of these countries and consequently on the functioning of aquatic ecosystems. It will force governments and communities to make choices in the ecosystems that should be protected and those that are too expensive, in water terms, to be maintained.

It is possible to develop transboundary water policy with an environmental component, but it is difficult. Transboundary water-sharing figures prominently in the South African water policy, although transboundary sharing is not specifically linked to ecosystem water needs. Even the Australian and Florida water policies have cross-border (interstate) concerns.[1] The Australian policy requires that environmental and other public benefits be achieved through joint arrangements for shared water resources, while the Florida water policy has interstate agreements with the states of Georgia and Alabama as one of its objectives, although this objective is not linked to environmental water allocations. It has taken substantial expenditure and national leadership to bring about the environmental water improvements across the eight Australian jurisdictions; progress has been considerably slower than envisaged when the water reforms commenced in 1994. In the EU, establishing a consistent ecosystem health policy across such a diverse region has proven to be expensive and time-consuming, but it may provide the basis for easier transboundary water management in the longer term. The Lesotho Highlands Water Project (case study 14) provides another example where the new environmental flows policy is part of the transboundary water agreement. The Senegal basin water charter (case study 16) and the Mekong basin agreement (case study 7) both contain specific provisions for environmental flows.

Environmental Water Mechanisms

The environmental objectives in a basin plan can be established using a variety of approaches. The South African and Tanzanian policies and legislation establish a national river classification system, where a national program is used to establish a quality class for each significant water resource. This class then acts as an environmental goal for the management of that water resource. Different water bodies will be assigned to different quality classes, depending on factors such as the uses of the water resource, its biodiversity values, and other factors. In the EU policy, by way of contrast, the environmental objectives for all water resources are the same, at least "good ecological status" (with some specific exceptions).

The Australian and Florida policies are different again in that neither introduces a national classification system. Instead, the environmental objectives are established as part of the catchment-level water allocation planning based on local, national, and international objectives.

All approaches have advantages and disadvantages. The EU approach, with its mandated uniform minimum environmental quality objective, minimizes the possibility that ecological outcomes will be diluted in the stakeholder negotiations, particularly where the river or groundwater system is already heavily committed with existing licenses. The Australian, Floridian, South African, and Tanzanian approaches are potentially more flexible and cost-effective, since not all water bodies have to be brought to the same environmental standard if competing water uses have higher priority.

Market mechanisms can be used for providing environmental water, but require an established infrastructure. While the South African and Tanzanian policies recognize the potential of using market mechanisms for water trading, only the Australian policy requires that market mechanisms be used for trading water allocations and entitlements for environmental purposes. However, even though there are functioning water markets in Australia, this component of the policy has been implemented quite slowly, primarily because of political limitations rather than institutional or legal barriers (Scanlon 2006). Large quantities of environmental water have now been bought on the water market by Australian governments, but there has yet to be active trading in environmental water allocations.

Once allocated, it is very difficult to recover water for the environment in overallocated systems. Recovery of overallocated systems receives particular emphasis in the Australian, EU, Floridian, and South African policies. However, the implementation of this requirement differs greatly. In South Africa, although there is yet to be any action to recover water for the environment, the water resource strategy acknowledges that about 50 percent of water management areas are currently overallocated. In Australia, where several catchments are clearly degraded because of lack of flows, there is little official acknowledgment of overallocation. Even so, many billions of dollars have been allocated to recovering water for the environment in stressed catchments in Australia. In Florida, too, there is little official acceptance of overallocation, with only one catchment being declared as overallocated.

It is clear from the experience to date in Australia, the EU, Florida, and South Africa that recovering water for the environment once it has been allocated to a consumptive water use is extremely difficult and politically unpopular. This has been identified as one of the major impediments to environmentally sustainable water management in both Australia and the EU (National Water Commission 2007). Even purchasing water at market prices from willing sellers in overallocated systems has proven unpopular in some parts of Australia. While legislative provisions for recovering water for the environment have yet to be used in South Africa, provisions in the National Water Act require new water licenses to be issued to replace existing entitlements. This will provide an opportunity to reduce current allocations and recover water for environmental purposes if systems are overallocated. The experience is similar in the EU. When the Guadiana River Basin Authority in Spain tried to reclaim allocated water resources, water users filed 15,000 separate court cases that stalled the process.[2]

The focus on recovery of overallocated systems can divert attention from protecting presently unstressed systems. The Australian and EU policies place considerable emphasis on the recovery of presently overallocated systems to environmentally sustainable levels of extraction. However, the experience in Florida is that this emphasis on stressed and potentially stressed water bodies means that more pristine systems are sometimes allowed to degrade substantially before any attention is placed on them. Consequently, the Florida policy requires the identification of water bodies that will potentially be stressed within 20 years and the development of a recovery or prevention strategy. This forward-looking provision thus focuses attention on the management of systems that need protection.

Participation

Participation is increasingly accepted as necessary even when its requirements in the policy are not very clear. The participation requirements vary considerably across the policies. The South African policy, legislation, and strategy have extensive stakeholder participation requirements, while Australia's and Florida's policies promote participation at specific stages of the development of water allocation plans. The European Union WFD has broad, timed requirements for public information and opportunities for public comment in river basin management planning, but no specific stakeholder involvement requirements in setting the environmental objectives or in undertaking the EFAs. In Tanzania, there is only a general requirement that concerned stakeholders be consulted when the national water resource management plan is drawn up; there are no explicit requirements for stakeholder consultations when the basin-level water resource management plans are developed.

In spite of this diversity of requirements, there has been considerable emphasis on stakeholder participation when the policy provisions were put into practice. For example, the first Tanzanian EFA undertaken in anticipation of a river basin water resource management plan (case study 8) has undertaken extensive stakeholder involvement activities in spite of the absence of direction in the policy. Similarly, the water management districts in Florida went well beyond the formal participation requirements when they established their water management plans.

Stakeholder participation can impede implementation unless carefully designed to suit a country's circumstances. However, participation can also act as an impediment to policy implementation if the policy requirements do not match the capacity of stakeholders or resources to undertake effective participation. Participatory requirements have resulted in delays in introducing environmental water provisions in South Africa and obstruction of efforts to recover from overuse in the case of the upper Guadiana basin in Spain (case study 2).

Assessment Method and Data

Provisions for "best-available science" in water policy can be used to impede policy implementation. Both the 1994 and 2004 policies in Australia require that best-available science should be used to assess environmental water needs. By way of contrast, the South African water policy does not mention whether the information used has to be the "best available" or not. In spite of these different policy requirements, both countries have been at the forefront of developing environmental flow assessment methods and applying them with high-quality scientific information. The Tanzanian policy also requires that "water for the environment shall be determined on the best scientific information available." The European Union WFD does not require any particular standard of scientific input to environmental water decisions, but it does require member states to give responsibility for achieving "good status" to a "competent" authority with the necessary scientific skills and mandate. Furthermore, there has been an extensive scientific effort across the EU to develop assessment procedures that are based on high-quality scientific information. Florida also requires that decisions be based on best-available information, but this has proven to be an impediment to progress in environmental water allocation because of concerns that decisions based on anything other than the highest-quality scientific information will be challenged in courts.

Value-laden terms in water policies need to be supplemented with specific interpretation and implementation mechanisms. All five policies contain value-laden terms, such as "significantly harmful to water resources or ecology" (Florida), "environmentally sustainable levels of extraction" (Australia), and "degraded beyond recovery" (South Africa) to describe water resources that are unacceptably environmentally stressed. However, these terms are very difficult to define operationally. To decide on what constitutes "significant harm," "sustainable levels of extraction," and "degraded beyond recovery" requires social decisions on the ecosystem services to be provided for different social groups. It would be difficult enough to make these social judgments if there were good evidence about the consequences of different decisions, but in reality often little information is available on the consequences of the different choices. For example, in the EU, scientists themselves do not agree on some biological, ecological, or hydrological issues. The European Union and Australia have mounted significant efforts to define these terms operationally.

Reviewing, Monitoring, and Enforcement

Establishing an independent oversight authority, with power to levy sanctions, can be an effective mechanism with which to implement environmental water provisions. The Australian National Water Initiative requires regular reviews of progress in implementing the initiative by the Australian state governments. A special authority—the National Water Commission—has been established to oversee the implementation and to undertake the reviews. For the first two years, it inherited the authority to recommend the withholding of payments to the state governments if inadequate progress was made with the broad range of water reform measures under the Australian national competition policy. This has proven to be a real incentive for compliance. The European Union WFD also requires each country to report on their progress in implementing the directive to the European Union. They can be fined for lack of compliance. Even though no fines have been levied yet, the provision has acted as a powerful lever for action by countries that see noncompliance as politically damaging.

These oversight mechanisms have been implemented in the two federated systems within the policy case studies. The individual country water policies do not include provisions for reviewing the implementation of the policies, although the Australian and EU experiences show the benefits of establishing reporting requirements and oversight of progress in implementation.

Environmental indicators need to be established, and the monitoring program should be focused on these outcomes rather than only on hydrologic measures. Both Australian and South African policies promote monitoring of water plan outcomes; that is, monitoring of environmental outcomes. South Africa is developing an ecosystem monitoring program, and various Australian states have developed programs, although their detail and focus on environmental outcomes vary greatly. The Florida legislation also requires annual reporting by both the water management districts and the Department of Environment Protection on establishing and meeting the minimum flows and levels. However, these reports have been restricted to hydrological measures and do not provide information on ecological outcomes.

Institutional Drivers

Environmental flows were just a component of broader water reforms in the five case studies. Thus the drivers consist of two parts: (a) drivers for the overall water policy reforms and (b) drivers specific to the inclusion of environmental flows in the policies (see table 5.2).

A singular event can be a powerful inducement to policy reform if professional and public drivers are organized to take advantage. The policy reforms in three of the five case studies were driven at least partly by drought, which focused attention on equitable water management. In Australia, Florida, and Tanzania, severe droughts highlighted the overallocation of water resources and the resulting stress on environmental functions. In Tanzania, the environmental aspects of this poor planning and management only became apparent later, as highly contested water rights issues gained high political visibility (see box 5.2). The South African case study was driven by a different singular event—the general reforms accompanying the democratic government in 1994—but also against a backdrop of growing water shortages. These unusual events, while difficult to predict and plan for, provide powerful stimuli for reforming water policies, with the concomitant opportunity to ensure that environmental sustainability and provisions for environmental flows are included in the new policies.

Public pressure can be a powerful driver for policy reform. Public pressure was central to both the water reforms and the inclusion of environmental flows in the new water policies in all cases except the EU Water Framework Directive. In South

---

BOX 5.2
Water Use Conflicts in Usangu Plains, Tanzania

Water shortages in the Great Ruaha basin, Tanzania, have resulted in intense competition between irrigators and pastoralists, particularly during the dry season. In the Usangu plains of the basin, water scarcity has caused tensions over access to both land and water. There was a perception among farmers that increasing numbers of cattle were placing greater demand on water and forage during the dry season, both within and around the Utengule swamps. At the same time, the gradual expansion of areas under irrigation by farmers decreased the amount of land that was previously available for grazing and the availability of water for livestock. The pastoralists and their cattle trespassed on cultivated fields to access water sources during the dry season, causing severe damage to the crops and cultivated fields and intensifying the hostility between farmers and pastoralists.

A DFID-supported study was initiated to obtain a scientifically credible explanation for the water shortages in the Great Ruaha basin. Its findings made it clear that livestock numbers in Usangu were smaller than previously claimed and that livestock water and pasture needs were within the carrying capacity of the basin. They were not the cause of either the water shortage or the environmental degradation within the basin.

WWF Tanzania, in close cooperation with the Rufiji Basin Water Office, is undertaking a study to identify and investigate options to restore flows to the Great Ruaha River flowing through the Ruaha National Park in Tanzania. The study will develop a short list of flow options and a high-level analysis of the preferred options. WWF has established an electronic forum to promote wide discussion of the prefeasibility assessment of potential options.

Sources: Ministry of Water and Livestock Development, Tanzania 2002; personal communication, Dr. Constantin von der Heyden, WWF Tanzania.

Africa, there was a strong pressure from the black majority to reform the water laws so that blacks would have equitable access to the country's water resources. Generally, they had much less interest in environmental flows, even though the services delivered by these flows were ostensibly to their benefit. In Australia and Florida, the public pressure included an awareness of environmental degradation and a desire to see a return of environmental values.

Scientific institutions can play a leading role in introducing environmental flows when policy reforms are occurring. This is seen most clearly in the South African reforms, where the scientific organizations were well organized and used the opportunity of the reforms to work with government officials to incorporate extensive environmental water provisions into the policy white paper.

International agreements that reflect an emerging consensus can be an effective motivator for the inclusion of environmental flows in national policies and legislation.

The global consensus that emerged in the early 1990s about the importance of environmental sustainability, including in water resources management (such as the Dublin Principles), was an important, albeit secondary, influence on the inclusion of environmental flows in the South African and Australian policies (the Florida policy and legislation predated this consensus by a decade). In Tanzania, there was a rising awareness among government officials about the weakness in the existing water resources planning and management decision-making process and about the emerging international consensus that included environmental water concerns.

Summary of Policy Lessons

The analysis of environmental flows in the five national water policies offers the following lessons:

- Countries in both developed and developing countries are integrating environmental flow provisions into their water resources policies.
- Gobal leaders in environmental flows include both developed (Australia, EU, and Florida) and developing (South Africa and Tanzania) countries.
- Some of the important aspects to be included in environmental flows provisions in policies are (a) legal recognition of environmental flows with, ideally, equal legal standing for consumptive water uses; (b) links between environmental flows and ecosystem services provided by the flows; (c) inclusion of all relevant parts of the water cycle, especially surface water and groundwater, when establishing environmental flow provisions; (d) a method for determining environmental objectives and outcomes at the basin level; (e) attention to both recovery of water for the environment in overallocated systems and protection of environmental flows in systems not yet under stress; (f) clear requirements for stakeholder participation in environmental flow decisions that do not impede progress; (g) the desirability of using an independent authority to audit performance of the policy; and (h) requirements for best-available science in making environmental water allocations, as long as this does not inhibit implementation of the policy requirements.
- Implementation challenges include (a) obtaining continued political support to implement the environmental flow provisions of the policy; (b) reorienting sectoral ministries to the need to include environmental water provisions in their policies and practices; (c) obtaining stakeholder support for environmental water provisions, especially in overallocated catchments and basins; (d) establishing environmental goals and the benefits delivered by associated ecosystem services; (e) turning value-laden terms such as "overallocation" and "sustainable levels of extraction" into practical procedures; and (f) matching the EFA procedures to the budget and time available, while still meeting the requirement for "best-available science."

TABLE 5.2
Institutional Drivers for Water Policy Reform and Inclusion of Environmental Flows in Policies

Country or Region	Policy initiation and implementation			Environmental flow inclusion[a]				
	Convening	Singularity	Public	Institutional	Evaluative	Public	Scientific Professional	International Developments
Australia	Federal government initiated national water reforms and provided financial assistance for their implementation	Drought accentuated the need for improved water management	Strong public pressure existed to make Australia's water management more efficient and environmentally sensitive	Professional water managers supported the inclusion of environmental flows in water policy	An independent organization was established to drive the reforms and recommend financial penalties for lack of progress	Strong public pressure existed to arrest the environmental decline of water resources	Scientific studies publicized the decline of environmental assets, and leading scientists advocated environmental flows	International consensus added support for environmental water reforms
European Union	EU was established to provide harmonized legislation and procedures across countries			The water supply and sanitation sector supported ecosystem health to reduce treatment costs; other water-using sectors supported ecosystem health because it could provide green credentials	The EU has power to sanction countries not meeting implementation standards		Ecosystem professionals advocated that ecosystem health was the best indicator of sustainable development	

Florida	Severe drought accentuated the need for improved water management	There was public pressure to manage water resources better in the face of the drought		Public and environmental groups advocated environmental restoration, especially of the Everglades	International consensus added support for environmental water reforms
South Africa	Installation of democratic government led to a policy to support redistribution of resources	Public pressure for water reforms was part of democratic change	Scientific organizations were active in advocating the inclusion of environmental flows in the policy		
Tanzania	Drought, inadequate investment, and poor water management led to electricity and food shortages	Public disquiet with the poor performance of the water sector, especially hydropower, contributed to the revision of the water policy	River basin officers and Department of Water Resources played a key role in the inclusion of environmental flows in policy	Supporters of the Ruaha National Park were very vocal about the drying up of the Ruaha River during dry seasons	International consensus added support for environmental water reforms, which were also influenced by water reforms in South Africa

Source: Authors.

a. Environmental flows were not specifically included in the European Union WFD. The drivers in this case refer to the policy's focus on ecosystem health.

Notes

1 Cross-border issues are those that cross water management boundaries within a country. They can share many characteristics with transboundary water management issues.

2 This example illustrates a related issue: the jurisdiction of the River Basin Authority may not match the area of the groundwater resources, particularly if they are deeper regional aquifers, that it has to manage.

CHAPTER 6

Basin Plan Case Studies: Lessons

FOUR CATCHMENT- OR BASIN-LEVEL water planning studies that included environmental flow assessments were selected for case studies (see box 6.1). They included one from a developed country (Australia) and three from developing countries and regions (South Africa, Tanzania, and the Mekong region). These case studies are described in Hirji and Davis (2009a).

Assessment of Effectiveness

As explained in chapter 4, the effectiveness of environmental flows in basin plans was assessed using the criteria of recognition, comprehensiveness, participation, method of assessment and data, integration, cost-effectiveness, and influence.

Recognition

Recognition of environmental flows in legislation and policy can simplify their incorporation into water allocation plans. Policy and legislation provide legitimacy for acceptance and implementation of environmental flow determinations. In the Kruger National Park, the instream flow requirements, first determined in the 1990s, were not implemented because they were not backed by legislative requirements and did not have mechanisms for implementation. However, once the

BOX 6.1
Basin-Level Environmental Flow Assessments

Kruger National Park, South Africa. There have been several studies of environmental water needs in the seven major rivers flowing through the Kruger National Park. These studies were initially driven by drought during the 1980s, when there was concern that the park's rivers were drying up because of upstream abstractions. Some of the EFA techniques then being developed in South Africa, such as BBM, were used to establish instream flow requirements. However, these IFRs were not implemented because there was no mechanism for allocating water to provide the park with a greater share. The advent of the National Water Policy and National Water Act in the late 1990s provided such a mechanism. The estimates of water required to support the park's ecosystems have been refined through more recent studies in order to provide estimates of the ecological reserve as defined in the act.

Mekong basin, Southeast Asia. The agreement between the countries of the lower Mekong basin includes provisions for establishing minimum flows and for maintaining the reverse flow to Tonle Sap in Cambodia. GEF and the World Bank helped the Mekong River Commission to implement these provisions through a three-stage process to explore the impacts of different environmental flow provisions. However, the environmental flow provision was reduced from the status of a rule to that of a guideline because there was a perception, even by the downstream countries, that it would impede development.

Pangani basin, Tanzania. The Pangani basin has experienced severe competition for water between the hydropower and irrigation sectors. Under the Tanzanian national water policy, each basin water office will be required to establish a water resources management plan, including provisions for environmental water allocations. The Pangani Basin Water Office and IUCN have commenced a trial EFA in the basin to explore the potential consequences of different flow scenarios. This has also been used as an opportunity to train staff from other basin water offices, as well as academics and ministry staff in EFA procedures.

Pioneer catchment, Australia. Under the 1994 COAG water agreement and the subsequent National Water Initiative, all Australian states are required to develop water allocation plans, including environmental flow provisions, in all significant surface water and groundwater systems. The Pioneer catchment water allocation plan was completed in 2002 as part of that commitment. The EFA for the plan was developed as a technical exercise using a holistic assessment technique, and the environmental water requirements were then incorporated into the final catchment plan. The plan includes a monitoring plan; early indications are that the environmental objectives are being met.

National Water Act was passed, these instream flow requirements (IFRs) were used to develop high-confidence reserve determinations for the rivers of the park, which will be incorporated into the catchment strategies to be drawn up by the catchment management authorities. Even so, there was still a lack of acceptance by upstream extractive water users about the limitations imposed on their taking of water. Similarly, in Tanzania, the Pangani basin trial EFA, the Mara catchment EFA, and the Wami basin EFA are being conducted with the knowledge that the new Tanzanian Water Resources Act will soon be passed. In the meantime, these EFAs and that for the Lower Kihansi Environmental Management Project (LKEMP) (case study 15) were legitimized by the national water policy, which requires environmental water allocations to be included in basin plans. There is, however, a danger that EFAs that were developed ahead of legislation may not meet the legal requirement for inclusion in a basin plan.

Following implementation, there is a need to demonstrate the benefits of environmental flows. The Pioneer catchment provides an example where environmental flows were incorporated into a water allocation plan under legislation—the Queensland Water Act 2000—and were accepted by all parties as a legitimate use of water. Even so, agricultural water users in the catchment needed to be reassured that the environmental flows would result in clear environmental outcomes, such as increases in native fish, healthy wetlands, and maintenance of estuarine mangroves. (This illustrates the importance of an environmental monitoring program, public reporting, and adaptive management.)

It is difficult to get agreement on transboundary environmental water planning. The Mekong basin provides the only example of basin planning for environmental flows within a transboundary setting.[1] It illustrates the tension between development aspirations and provision of ecosystem services when each nation is focused on sharing water rather than sharing benefits. Even though the Mekong agreement contains requirements for minimum flows and maintenance of the reverse flows to Tonle Sap, the concept of environmental flows has not been fully accepted by all basin countries because of its perceived restrictions on development. Nevertheless, attempting to resolve transboundary environmental flow issues using formal EFA methods may contribute to the resolution of wider transboundary issues.

Comprehensiveness

All relevant components of the water cycle should be considered in the EFA. The Pangani basin and Pioneer catchment EFAs illustrate the hydrological integration of estuarine and freshwater needs for environmental flows. Both also demonstrate the consideration of ecosystem services from both surface water and groundwater. The Pangani basin EFA is the only one of the four basin-level case studies to include the effects of climate change in its scenarios. All but one of the scenarios assessed included the best estimate of the effects of climate change on the basin's water resources.

Participation

Participation is important but needs to be realistic; it should be tailored to suit the capacities of the stakeholders and the policies of the country and to build capacity in IWRM. Local communities did not have a history of participating in decisions about resource use in the catchments of the Kruger National Park. Although they were keen to improve their access to water, they lacked the institutional forum, and perhaps the confidence, to work with the predominantly white constituencies and lacked the capacity to engage fully in discussions. In the Mekong basin, it was difficult to develop a full stakeholder engagement program, when there were such strong differences in government attitudes, stakeholder capacities, and language. The initial EFAs were undertaken as technical assessments with limited stakeholder engagement, while an inclusive study is yet to be fully implemented because of funding limitations and lack of comprehensive support from the basin countries. In both the original IFR studies and the more recent reserve determinations in the Kruger National Park, there was only limited direct engagement by local groups, although the impacts of flow decisions on their livelihoods were considered by the planners. Experience suggests that it takes time to develop the capacity for stakeholders to participate effectively in activities such as EFA. Specific capacity-building activities are currently under way, including development of a common vision, establishment of local objectives, and stakeholder involvement in monitoring activities.

Publication of submissions and government responses can assist transparency in decision making. The Pioneer catchment (case study 9) illustrates an effective mechanism for promoting transparency. Under the Queensland Water Act 2000, all submissions to the water planning process have to be published, together with the government's response to each submission, within 30 days of the plan being approved.

Stakeholder engagement in transboundary settings is particularly difficult. This is illustrated by the Mekong basin (case study 7). Not only did the different stakeholders within the basin have different languages, capacities, and objectives, but the different governments had very different attitudes toward the involvement of local groups in national development decisions.

Assessment Method and Data

For both plan and project EFAs, a range of techniques are needed within a country to meet different levels of environmental risk and to suit different budgets and time frames. All four basin or catchment case studies used versions of the holistic approaches to developing EFAs, based on both general flow–ecology relationships and specialized field studies. The building block method (BBM) was used initially in the Kruger National Park low flows, followed by the flow stress ranking method. The benchmarking method was employed in the Pioneer catchment, and modified versions of the DRIFT (downstream response to imposed flow transformation) method were used in the Pangani basin and the third phase of the Mekong basin. In the case of the Pangani basin, experienced international consultants devised a procedure that assessed the flow requirements of the major ecosystem components.

The experience in both Australia and South Africa is that a country needs to implement a range of EFA techniques for incorporating EFAs in basin-level plans to suit the budget, skill, and information needs as well as the severity of the pressure on the environment. Thus South Africa has adopted four levels of EFA procedure (see box 6.2), while Australian states such as New South Wales have two basic EFA methods, each being applied with more or less intensity to suit the circumstances.

Similarly, in the project case studies, the assessment methods varied from the simple estimates of restored lake levels that were undertaken for the Aral Sea case study, to the hydrological and hydraulic modeling undertaken for the Chilika lagoon study, to the detailed and expensive flow assessment method (DRIFT) undertaken for the Lesotho Highlands Water Project (LHWP). The high cost of DRIFT (nearly $2 million) and the time needed (more than two years) were justified because it was important to have defensible and comprehensive results for a very large project ($2.9 billion) that were grounded in specific impacts to convince skeptical managers in the Lesotho Highlands Development Authority (LHDA).

BOX 6.2
Levels of Environmental Flow Analysis Used in South Africa

The South African Department of Water Affairs and Forestry has developed four levels for determining environmental water needs: desktop, rapid, intermediate, or comprehensive. The method used depends on the environmental pressure faced by the body of water and the funds and time available. The desktop and rapid determinations are based largely on applications of the BBM to determine instream flow requirements (King and Tharme 1994; King, Tharme, and de Villiers 2000; Tharme and King 1998). The intermediate and comprehensive determinations, which can be based on the BBM, DRIFT, or flow stress ranking methods, involve specific local data collection and hydraulic modeling.

Source: Personal communication, D. Louw, March 2008.

Where possible, field data should be used to supplement desktop assessments. The collection of field data was an important component of all planning studies in order to establish defensible flow-ecology relationships. While much of the data collected under the Kruger National Park Rivers Research Program were not directly usable in the EFAs, the understanding and knowledge of the scientists engaged in the program were a valuable resource for the EFAs. The Pioneer catchment EFA relied on two years of field data assessments to illustrate the risks to different organisms from the two flow scenarios examined, while the Pangani basin study undertook data collection in the freshwater and estuarine reaches of the basin. The second and third phases of the Mekong basin study also used field data.

An ecological monitoring program is a key part of a basin plan. Establishing an environmental monitoring program is an important, but often neglected, part of implementing environmental flows. An ecological reserve monitoring program is being developed for some catchments of Kruger National Park, but, of the case studies, only the Pioneer catchment (case study 9) has advanced to the stage where such a monitoring and reporting program is operational. In this plan, five environmental assets have been identified, and annual reports are being produced on the delivery of environmental flows and the state of these assets. Under the Queensland Water Act 2000, the minister for water is required to prepare a regular report outlining progress on the implementation of a water resources plan and the achievement of the plan's objectives. These monitoring and reporting requirements not only provide feedback about the success of the measures in the plan, but also provide a public driver for the continued attention by government to providing environmental flows.

Integration

For both basin plans and projects, environmental outcomes can be integrated with social and economic outcomes either as part of the EFA process or during the decision-making process. The case studies illustrate two approaches to integrating the environmental assessment with social and economic issues. The Pioneer catchment EFA dealt only with environmental water needs and did not explicitly integrate these needs with social or environmental uses of the water. These environmental water needs were then traded off against other demands on the catchment's water resources during the water allocation planning process. Similarly, the Bridge River consultative committee (case study 12) used both intuitive and formal methods to combine the environmental, social, and economic outcomes for different flow scenarios in the Bridge River assessment. One consequence of this approach is that, without a very transparent decision-making process, it is difficult to assess the extent to which the environmental flows in the final plan protect the environmental assets and ecosystem functions in the catchment.

The Pangani basin EFA and the LHWP illustrate an alternative approach. The assessment of different environmental flow scenarios included the social and economic benefits to the communities that depend directly on the river flows. In a similar way, the second and third phases of the Mekong basin study progressively included social, economic, and environmental benefits from the different environmental flow scenarios. In that case, it was clear that an analysis focused just on environmental outcomes would not have been accepted.

Cost-Effectiveness

Little information is available about the cost of the basin and catchment EFAs. The Pangani EFA trial cost approximately $500,000 over three years, but this figure includes the training and establishment costs that would not be incurred in subsequent applications in Tanzania. Nevertheless, an EFA that included the extent of fieldwork included in these EFAs would be expensive, even though these costs are a small fraction of the benefits obtained from development. However, a lower-cost approach might be more appropriate for catchments in developing countries where the development pressures are not yet intense.

The Pioneer catchment EFA was one of the first EFAs carried out in Queensland as part of the development of catchment water allocation plans across Australia. The assessment was judged to be too costly for widespread application if the intensity of scientific investigation remained at the level employed in this catchment. Consequently, there has been a rationalization of EFAs, with less-intensive approaches being employed in Queensland catchments, where the ecological risks are judged to be lower.

Influence

The Pioneer, Kruger National Park, and Pangani EFAs have all been influential in different ways. The Pioneer EFA results fed directly into the water allocation planning of the Pioneer catchment; the environmental assets now receive environmental flows, although the extent to which the flows in the EFA are actually delivered is difficult to assess. This work has not had influence outside the catchment because there was already a general program to roll out water allocation plans across the country.

The extensive environmental flow investigations in the Kruger National Park have been highly influential, both within the catchments surrounding the park and more widely. The early IFR determinations formed the basis of the subsequent ecological reserves, although these determinations have yet to be formalized in water allocation plans for the park's catchments. The EFAs also gave confidence to those developing the South African water policy and legislation that it was possible to estimate the flows needed to maintain aquatic environments.

While the trial Pangani EFA has yet to be completed and the basin water resource management plan has yet to commence, the study has already been influential in both training a cadre of academics and government staff in the practical aspects of EFA (with some trained staff now participating in EFAs in other basins) and in increasing awareness of environmental flow issues in this basin.

It is more difficult to mount an influential EFA in transboundary water management. The EFA in the Mekong basin, the only fully transboundary EFA in the plan case studies, has produced some valuable documentation on basin hydrology and has brought about some changes in attitudes toward environmental water provisions. However, Mekong basin countries are yet to fully embrace the concept of environmental flows. The minimum flow rules that were recommended as a result of the initial EFA were subsequently reduced in status to guidelines, and the third phase of the EFA has been hampered by a lack of support at senior political and administrative levels.

Institutional Drivers

Table 6.1 shows the institutional drivers that operated in the four basin or catchment case studies.

Procedural drivers were important motivators for all four planning case studies. However, the details differ about the roles of these drivers. The Pioneer catchment EFA was a direct response to the legal requirement for water allocation plans to be developed for all major surface water and groundwater catchments in Australia and was accepted by all sectors without issue. The Mekong EFA was undertaken in response to the Mekong agreement. The Pangani EFA was undertaken in anticipation of passage of the Tanzanian water legislation; the trial would probably not have occurred in this basin if the IUCN had not acted as an additional driver because of its interest in locating a suitable basin in which to conduct an EFA demonstration study. The Kruger National Park EFAs were only driven by procedural requirements after passage of the 1998 National Water Act; prior to this, the professional and public drivers had not been sufficiently strong to lead to agreed allocations by the Department of Water Affairs and Forestry of water for maintaining the park's ecosystems.

However, other factors, apart from these institutional drivers, were important in initiating some of the case studies. The concern among some of the countries of the Mekong basin about proposed developments in the upstream China tributaries of the Mekong River was a powerful impetus for seeking rules for sharing the waters of the basin. GEF and the World Bank provided support for an EFA to implement some of the provisions of the Mekong agreement.

TABLE 6.1
Institutional Drivers for Undertaking Environmental Flow Assessments at Basin and Catchment Scales

Site	Procedural	Evaluative	Instrumental	Professional	Public
Kruger National Park	The 1998 National Water Act provided the justification for establishment of the ecological reserve in the park's rivers			Park managers and scientists were concerned about the effects of water abstractions and proposed dams on the park's biodiversity	NGOs were concerned about the impacts of dam proposals on the park's biodiversity
Mekong basin	The Mekong agreement requires protection of low flows and the reversal of flows to Tonle Sap		The World Bank and GEF included EFAs as part of their development assistance		International, regional, and national NGOs raised concerns about development proposals on the Mekong River
Pangani basin	Tanzanian national water policy and draft Water Resources Act require EFAs		The IUCN supported the EFA trial as part of its Water and Nature Initiative		
Pioneer catchment	The 1994 COAG agreement and the 2000 Queensland Water Act required EFAs as part of catchment-level water allocation plans	The National Competition Council and the National Water Commission reviews of progress with implementation of the National Water Initiative provided an evaluative driver for the EFA		The government water managers supported the formal allocation of water for the environment to prevent the catchment from becoming stressed	Public opinion was strongly in favor of environmental sustainability, both in the Pioneer catchment and more generally

Source: Authors.

Summary of Plan Lessons

The following are some of the key lessons to emerge from this analysis of four catchment and basin plans:

- It is simpler to implement environmental water provisions in basin and catchment plans when the concept has been recognized in policy and legislation.
- It is not enough to allocate water for the environment in basin plans; managers need to demonstrate the resulting human and ecosystem benefits through monitoring and interpretation.
- Caution is needed when allocating water rights; once allocated, it is very difficult to return water to the environment.
- Participation needs to be tailored to meet the capacity of the stakeholders to engage in decision making; this should include mechanisms to help them to understand the issues and consequences of decisions.
- No single EFA technique suits all basin planning occasions; a range of techniques from the simple to the complex is needed to respond to the different levels of risk and intensity of water use.
- Ecological monitoring of the outcomes of the plan is essential, partly to reassure stakeholders that the environmental benefits are being delivered and partly to provide information for adaptive management.
  In addition, the case studies highlight challenges related to implementation:
- Incorporating water-using land-use activities, such as plantation forestry, in basin water allocation plans remains difficult.
- Developing water allocation plans that take account of all downstream water needs, including estuarine, near-shore, and groundwater needs, is difficult partly because there is a lack of understanding about the dependence of these systems on freshwater flows and partly because the institutions that manage these components of the water cycle are often separate from the institutions that manage surface freshwaters.
- Building the expertise and data to undertake basin plans with environmental flow components remains a major challenge in developing countries.
- There is a perception that undertaking EFAs is expensive and that the water that is allocated for the environment would otherwise be used for productive activities. This perception will continue, until credible cost-benefit analyses are conducted to quantify the benefits and losses from different allocation decisions.

Note

1 Although the Pangani basin and the catchments of the Kruger National Park are transboundary, the EFAs in these cases were carried out in just one country.

C H A P T E R   7

Project Case Studies: Lessons

THE PROJECT-LEVEL CASE STUDIES INCLUDE EFAS conducted as part of the development of a new dam (Berg River Dam in South Africa and Mohale Dam in Lesotho), replacement of old infrastructure (Naraj Barrage on the Mahanadi River in India and irrigation canals in the Tarim basin in China), reconstruction or modification of existing infrastructure (Berg Strait Dike in the Aral Sea; lower Kihansi in Tanzania; and Katse Dam in Lesotho), and reoperations for existing infrastructure (Bridge River in Canada; Manantali Dam in the Senegal basin; dams on the Syr Darya River in the Aral Sea basin and the Tarim basin in China). Table 7.1 summarizes the characteristics of the case studies, and table 7.2 summarizes the major findings from the assessment. The eight project case studies are described in Hirji and Davis (2009a).

Some of the lessons for the rehabilitation and reoperation case studies differ from those for new infrastructure. Two of the major infrastructure rehabilitation and ecosystem restoration case studies—the Tarim basin in China and the Aral Sea in Central Asia—involved inefficient infrastructure, so there were opportunities to improve efficiency and redistribute the "saved" water to the environment as well as to achieve increased production. At the Manantali Dam in the Senegal basin, the hydropower turbines had not been installed for a decade after the dam was built, providing an opportunity to demonstrate through experimental flood release studies the value of environmental flows to downstream

TABLE 7.1
Characteristics of Project Case Studies

Case Study	Country or Region	Gross Domestic Product per Capita (US$)[a]	Institutional Setting	Sector	Purpose	Date Completed
Aral Sea	Central Asia	$260–$43,000	Transboundary	Environmental restoration	Reoperations and restoration; dike upgrading	GEF project 2003; World Bank project still active
Berg River	South Africa	$5,390	Catchment	Water supply	New dam, operating rules	In progress
Bridge River	Canada	$36,170	Subcatchment	Hydropower	Reoperations	2001
Chilika lagoon	India	$820	Subcatchment	Irrigation flood control	Restoration and reoperations	2004
LHWP	Lesotho	$1,030	Transboundary	Interbasin transfer (water supply)	Reconstruction of outlet structure in old dam and in new dam; new flow release policy	2006
Lower Kihansi Power Project	Tanzania	$350	Subcatchment	Hydropower	Reconstruction of outlet structure in new dam, spray augmentation using artificial sprinklers	In progress
Senegal basin	West Africa	$750	Transboundary	Multipurpose	Reoperation and restoration	Regional Hydropower Development Project completed 2005
Tarim basin	China	$2,010	Sub-basin	Irrigation	Irrigation canal reconstruction and reoperations	2005

Source: Authors.
a. From World Bank Doing Business 2008 site. Available at http://www.doingbusiness.org/ExploreEconomies/EconomyCharacteristics.aspx.

TABLE 7.2
Major Findings from Project Case Studies

Project	Recognition	Comprehensiveness	Participation	Assessment Method	Integration	Cost-effectiveness	Influence
Aral Sea	The water needed to restore the northern Aral Sea was unquestioned by local communities and government, although not called "environmental flows"	Flow needs were those required both to refill the northern Aral Sea and to reduce floods in upstream areas; however, the hydrograph was not decomposed into components	Success of the northern Aral Sea recovery was partly due to strong engagement by local Aral Sea communities and government	No EFA or quantitative modeling was carried out, apart from simple estimates of water balance and levels in the northern Aral Sea	Social and economic benefits were integral to project and not quantified in advance		Refilling of the northern Aral Sea reestablished the fishing industry; Lake Sudoche recovery raised interest in other lake restorations
Berg River	After initial reluctance, environmental flows have been strongly supported by government since the National Water Act was passed; community support was strong for environmental flows	The study considered a range of flow components, including the estuary (although this was not proceeded with)		Initial EFA was based on BBM, subsequently extended with field studies; an extensive monitoring program was integrated with adaptive management	EFA was fully integrated with the EIA during project preparation	Capital expenditure: $6.6 million to $14.9 million	The EFA has been influential in determining the initial operating rules, with further adaptation proposed as data arrive from the monitoring program; it is yet to be influential outside the Berg River

(continued)

TABLE 7.2
Major Findings from Project Case Studies (continued)

Project	Recognition	Comprehensiveness	Participation	Assessment Method	Integration	Cost-effectiveness	Influence
Bridge River	The dam operating authority preferred to maintain the agreed downstream releases, until threat of a court case persuaded them to investigate the downstream ecosystem water needs; there was strong community acceptance of the concept of environmental flows for specific species protection	Originally the project only considered minimum flows; subsequent study investigated the flows needed for a range of organisms	A consultative committee drove the process; there was strong engagement and ownership from all sectors, except the First Nations	A systematic method based on multicriteria decision making was used; the modeling results were couched in terms of agreed indicators that all decision makers understood; an environmental monitoring program was integrated with adaptive management	Separate technical studies were integrated intuitively and formally during decision making; flow modeling was combined with water quality modeling	EFA cost: $600,000 Monitoring and implementation: $520,000 per year	More power was generated, and better environmental outcomes were achieved; the project influenced other water use plans
Chilika lagoon	The successful restoration of the lagoon through engineering works blunted interest in establishing environmental flows from Naraj Barrage	Assessment focused on environmental flows to the lagoon only, but included some water quality aspects	Institutional stakeholders had little engagement in the environmental flow determination because the engineering work had been successful for the short term; community stakeholders were consulted as much as their capacity permitted	Hydraulic and hydrologic modeling combined with ecological outcomes, although formal EFA method was not followed; lagoon recovery is monitored, but not linked to flows from barrage	Flow modeling combined with water quality modeling		Draft environmental flow rules developed for the Naraj Barrage, although yet to be implemented, have led to the incorporation of environmental flows as a priority water use in the state water policy

Lower Kihansi gorge	Water resource institutions rather than environmental institutions took the lead with environmental flows. The dam operating authority was reluctant to provide water for downstream ecosystem; Tanzanian government organizations understood the relevance of the concept and insisted on provision of flows	Flow needs restricted to the gorge ecosystem but focused on unique mechanism of spray dependence	No downstream communities; the concerns of the international environmental groups were represented through the Tanzanian environmental organizations	Extensive fieldwork and experimentation to link flows through gorge with the extent of spray and ecological response. No formal EFA method was applicable; monitoring was a critical component, used for enforcement	No economic benefits; social benefits were integral to outcomes, but not quantified	The project stabilized the lower gorge, improved knowledge of environmental flows within government, and generated interest in catchment-level environmental water plans
Lesotho highlands	Managers initially persisted with an agreed minimum flow	The initial concept was of minimum flows; subsequent studies using DRIFT were fully comprehensive for in-channel and floodplain watering	Downstream communities were consulted through the DRIFT process, but their influence on decisions was quite limited	The DRIFT method was developed for this project, as was a method for presenting complex results	The DRIFT technique integrates environmental, social, and economic outcomes; EFA was meant to be integrated with EIA, but was delayed; flow modeling was combined with water quality modeling	EFA: $2 million; compensation: $14 million	Downstream environmental health targets were met or exceeded; DRIFT technique is being used more widely

TABLE 7.2
Major Findings from Project Case Studies (continued)

Project	Recognition	Comprehensiveness	Participation	Assessment Method	Integration	Cost-effectiveness	Influence
Senegal basin	Managers were initially reluctant to accept the concept, but attitudes changed once the water charter was signed	The project includes estuarine delta water needs as well as floodplain and in-channel needs	Downstream communities were involved with the assistance of an NGO	Hydrological models were used to predict the extent of floodplain inundation; no ecological monitoring was conducted	Separate economic and social studies were integrated during decision making with environmental and hydrological modeling of groundwater and surface water studies		
Tarim basin	Water needed to restore the green corridor was unquestioned by government because restoration was their priority	The flows were confined to those needed to reestablish the green corridor; the project included groundwater and surface water modeling of irrigation districts	Water user associations and irrigation district committees were formed and consulted, but were not drivers	No specific EFA technique was used, but hydrologic and hydrogeological models were used to predict water savings; monitoring of water use and downstream ecological response was integral to the project	Economic and social benefits were integral to the project, but EFA was not carried out; groundwater and surface water both were modeled but were not integrated	The increased agricultural production was greater than the cost of water efficiency work	The project achieved increased cropping and improved environmental outcomes

Source: Authors.

communities in time for the needs of their residents to be included in the final water-sharing arrangements.

Once rights to water from infrastructure developments have been established, it is usually difficult to redistribute them through rehabilitation projects. This is most apparent in countries, such as Australia and South Africa, that are facing major costs to return water to the environment in overallocated systems.[1] Not all of the costs are financial. In Australia, the political costs of buying back water licenses from irrigators are probably even greater than the financial costs. In Tanzania, the government faced significant political costs because the water being reserved for the unique ecosystem in the lower Kihansi gorge meant that electricity generation could not be expanded further in a nation facing enormous electricity shortages and high demand.

For restoration of degraded downstream ecosystems, engineering improvements are often needed to provide the volume of flows needed. To restore the northern Aral Sea, upstream dam-operating structures had to be modified to better control water releases, and a dike was constructed across the Berg Strait to retain the sea's water. Similarly, Chilika lagoon in India, Kihansi Dam in Tanzania, and the Tarim basin in China all required engineering interventions (respectively, a new entrance to the ocean and a channel across the lagoon, a modified outlet structure from the Kihansi Dam and an artificial sprinkler system, and a lining for irrigation channels to conserve water) along with the establishment of environmental flows for the recovery of downstream ecosystems.

Assessment of Effectiveness

As described in chapter 4, the effectiveness with which environmental flows were included in water resources projects was assessed against recognition, comprehensiveness, participation, method of assessment and data, integration, cost-effectiveness, and influence.

Recognition

Procedural drivers can help water resources managers to accept the concepts of environmental flows. The intuitive acceptance of environmental flows in the Aral Sea and Tarim basin case studies (see box 7.1)—the two ecosystems that have been most heavily degraded out of these case studies—can be contrasted with the initial reluctance to accept the concept in the Lesotho highlands, Chilika lagoon, lower Kihansi, Bridge River, and Senegal basin cases. In these cases, the operating authorities had a mandate to develop and operate infrastructure. However, in all these cases, with the exception of Chilika lagoon, the operating authorities came to understand and accept the relevance and legitimacy of environmental flow concepts. For the lower Kihansi and Senegal River cases, this acceptance was assisted by the

---

BOX 7.1
The Tarim Basin Restoration

The Tarim River, in western China, has been increasingly used for irrigated agriculture over the last 50 years—to the point where, in the 1970s, it ceased to flow in its lower reaches. Lake Taitema, a terminal lake, has not received water from the river for many years. The "green corridor" in the 300 kilometers of the river above Lake Taitema has became ecologically stressed, reducing the vegetative barrier to encroachment of the Taklamakan and Kukule deserts, which border the river in that area. The advance of the deserts and the threat that they would join up and sever the transport links was a significant concern to the national government.

   The Bank-supported Tarim II Project used geomembranes to line the leaky water distribution channels and significantly reduce water losses, together with drainage improvements. The saved water was used both to improve agricultural production in the irrigation districts and to return water to the lower reaches of the Tarim River. The project also improved management arrangements for the basin's water by establishing the Tarim Basin Water Resources Commission. The commission made it clear that it would enforce the quotas established for irrigation abstractions.

   As a result of the project, canal seepage losses have been reduced by between 600 million and 800 million cubic meters per year; 41,000 hectares of new irrigated land has been developed; the incomes of farmers have risen substantially; Lake Taitema has expanded to 200 square kilometers; and riverine vegetation has improved dramatically.

Sources: Hou and others 2006; World Bank 1998.

---

procedural drivers of a new policy and a new transboundary agreement, respectively; for the Bridge River and Berg River cases, the acceptance came about because the benefits of providing downstream flows became more evident as the EFA studies progressed. In fact, in the Bridge River case, the reoperation procedures resulted in increased power generation as well as improved downstream benefits through changes in the timing of water releases.

   The concept of environmental flows has never been fully accepted by the Water Resources Department and other Orissa State government departments, partly because the benefits to the Chilika lagoon were more distant than the immediate benefits from opening a new mouth to the ocean and partly because high staff turnover prevented the development of a corporate approach. The World Bank facilitated the EFA, but this was completed after the Bank project had closed. Thus the Bank was unable to influence the state government further to ensure that its recommendations were incorporated into the Naraj Barrage operating rules.

   Catchment-level water allocation plans provide benchmarks for project-level decisions on water allocations. Difficult decisions about sharing water from infrastructure development can be assisted if water policies and laws are in place that define environmental uses as legitimate uses of water that are protected by the force of law or if a water allocation plan is in place. Such a plan establishes the agreed distribution of water or benefits from river flows and acts as a benchmark for redistributing benefits from infrastructure. The Lesotho Highlands Water Project would have been simplified if such a plan had been in existence before the dams were developed.

Environmental flow provisions need to be monitored and enforced. Enforcing environmental flow provisions, like all allocations, requires vigilance. While the LHDA has formally agreed to the environmental flow provisions, there has been a lack of understanding and acceptance by operational managers and staff, and the agreed environmental flows have not always been released in a timely manner nor in adequate quantities; for example, important flood releases were not made. Similarly, diligent monitoring by the Rufiji Basin Water Office showed that the environmental flows from the lower Kihansi Dam during the first two years of operation were about 30 percent less than the flows that the power utility reported it was providing. This became evident after the monitoring system became operational.

Comprehensiveness

The concept that environmental flows are a matter of retaining minimum flows in rivers and estuaries arose in a number of case studies. Examples include the treaty minimum flow, which was agreed to in the 1968 treaty to build the Lesotho Highlands Water Scheme; the inclusion of minimum flows in the Mekong agreement; the 1998 proposal from BC Hydro to the Department of Fisheries and Oceans to provide minimum flows; and the Senegal basin water charter, which specified "minimum flows and other ecosystem services."

However, the EFAs carried out in these and other cases assessed all components of the flow regime and often recommended that some of the flow components (for example, dry- and wet-season flows, freshets, and occasional large floods) be retained to provide a range of downstream ecosystem services.

EFA studies should consider all downstream-dependent ecosystems. Several case studies included consideration of flows to maintain ecosystem services in estuaries and groundwater systems as well as surface water. The Chilika EFA focused principally on flows to the Chilika lagoon. The original environmental flow plans of the Organisation pour la Mise en Valeur du Fleuve Sénégal (OMVS) in the Senegal basin did not include provisions for freshwater flows into the delta and estuary of the Senegal River because subsistence farmers on the floodplains were not seen as contributing to the national economies of the basin countries. Once the importance of these flows was realized, the OMVS approved a water charter (see box 7.2), which included the delivery of water to these downstream environmentally sensitive areas as well as the provision of flood flows for the mid-river floodplains and groundwater recharge in the floodplain. The Berg River EFA

BOX 7.2
The Senegal Basin Water Charter

In May 2002, the governments of Mali, Mauritania, and Senegal signed a water charter. Guinea signed the charter in 2004. The objective of the charter is to "provide for efficient allocation of the waters of the Senegal River among many different sectors, such as domestic uses, urban and water supply, irrigation and agriculture, hydropower production, navigation, [and] fisheries, while paying attention to minimum stream flows and other ecosystem services." The charter guarantees an annual artificial flooding (Article 14) and minimal environmental flows (Article 6), except under extraordinary circumstances.

The charter contains the principles and procedures for the allocation of water and establishes a permanent water commission to serve as an advisory body to the OMVS's Council of Ministers.

The water charter extended stakeholder involvement within the Senegal basin to include farmers and NGOs. Further stakeholder participation was stimulated by the GEF project, which included participation in its design and implementation. Now local coordination committees exist throughout all countries of the basin.

Source: World Bank 2006a.

---

considered the need for flows to maintain the river's estuary, but decided that this was not an important aspect at that time. It has subsequently carried out detailed work on the estuarine reserve. Only the Tarim basin case study explicitly considered the effects that returning water to the environment would have on groundwater. In the Senegal case, the groundwater recharge was a secondary benefit.

None of these project EFAs considered the effects of climate change on the flows needed to maintain downstream ecosystem services.

Participation

While the involvement of stakeholders is always central to achieving successful environmental flow outcomes, the methods employed and the responsibility assigned need to be tailored to suit local circumstances. The Bridge River EFA was at one extreme, where the consultative committee drove the EFA process, with BC Hydro providing the secretariat. At the other extreme, the irrigators in the Tarim basin were involved through their water user associations and irrigation district committees, but the real decision-making authority resided with government departments and the national government. However, both projects resulted in successful environmental and production outcomes. The restoration of the northern Aral Sea provides another instructive example. Although the World Bank project included a component for a basin consultative group in the Syr Darya Control and Northern Aral

Sea Project, the group was never formed because of the difficulties in developing this group across the five countries of the basin. Nevertheless, the communities directly affected by the shrinking of the northern Aral Sea were highly motivated to engage in the project and provide assistance.

Stakeholder engagement mechanisms need to be designed around the capacity of communities to engage in decision making. The project where the EFA studies have not yet been influential—the restoration of Chilika lagoon—was an example where institutional stakeholders were not effectively engaged with the process. The stakeholder executive committee was not formed until late in the project, and the institutions, generally, did not engage in the EFA process. Nevertheless, local communities around the lagoon were consulted and did provide feedback of value to the EFA team, although their ability to comment on the technical options was very limited. Similarly, the affected communities in the Lesotho highlands and the Senegal River valley were limited by their understanding of the technical aspects of the assessment. In the former case, the lack of policy in Lesotho hampered the participatory process: there was no guidance on the role expected of the communities or the extent of consultation. There was a tendency for the LHDA to inform these communities rather than listen to their requirements.

Finally, the lower Kihansi EFA provides a unique example in which no downstream communities were directly affected by the change in river flows. The affected stakeholders were the international community. Through the Biodiversity Convention, they expressed the objective of protecting endangered species and ecosystems. These stakeholders were represented in the decision making by the appropriate river basin and environmental organizations and also by international NGOs, which had originally raised the alarm over the loss of the downstream ecosystem.

Assessment Method and Data

Information is of little value if it is not couched in terms that the audience can understand. One lesson emerging from the Lesotho highlands project is that the results of the scientific studies within the EFA must be comprehensible to managers who must determine the allocation of water to its different uses. In that example, the scientists undertaking the EFA devised a simple, understandable system for conveying the downstream consequences of different flow scenarios. Similarly, the information provided to stakeholders needs to be in a form that is understandable. In the case of the Berg River EFA, the language used by ecologists and other scientists was initially found to be a barrier to understanding the implications of different options. For this reason, the results of the scientific investigations in the Bridge River studies were couched in terms of the performance indicators for seven outcomes that had been agreed to by the consultative committee (see box 7.3).

BOX 7.3

Structured Assessment for the Bridge River Reoperation, Canada

The members on the consultative committee charged with developing the reoperation plan for the dams on the Bridge River came from a wide range of backgrounds. In order to work together effectively, they followed a systematic approach that consisted of six key steps and was based on multiple-attribute techniques and value-focused thinking (Keeney 1992).

The steps began with clear articulation of objectives. Performance measures were developed that describe the extent to which each alternative operating regime contributed to or detracted from each objective. Usually quantitative, the performance measures force specificity to the objectives, better educate each participant on the needs of others, and create a basis on which to collect decision-focused information.

The following objectives were agreed for the Bridge River water-use plan:

- Fisheries: maximize the abundance and diversity of fish
- Wildlife: maximize the area and productivity of wetland and riparian habitat
- Recreation and tourism: maximize the quality of the recreation and tourism experience
- Power: maximize the value of the power produced
- Flood management: minimize adverse effects of flooding on personal safety or property
- Dam safety: ensure that facility operations meet requirements of BC Hydro's Dam Safety Program
- Water supply and quality: preserve access to and maintain the quality of water.

In total, more than 20 alternatives were run through BC Hydro's operations model. The consequences for each objective were discussed by the consultative committee against the agreed performance measures. Preferences and values were documented, and areas of agreement were sought. The consultative committee members eventually agreed on a single recommended operating alternative.

Source: Bridge River water-use plan (case study 12).

Environmental flow monitoring programs need to assess the ecological outcomes, not just the flows themselves. Only some of the EFAs included monitoring components. Whether for rehabilitation or restoration cases or for new infrastructure, monitoring programs should be focused on ecological and social outcomes. Thus the Bridge River monitoring program assesses whether the downstream ecological outcomes (such as fish recovery) are achieved; the monitoring program in the Tarim basin has assessed not only the adherence to the water abstraction quotas, but also the extent to which the downstream riparian areas have recovered; the lower Kihansi monitoring program checks not only whether the agreed bypass flows have been provided but also the extent to which the gorge ecosystem has recovered; and a monitoring program has been established in the Berg River since 2002 to provide a baseline against which the effects of the new dam can be measured (see box 7.4). However, the Senegal basin does not appear to have an ecological monitoring program; the Chilka lagoon monitoring program is not designed to distinguish the effects of any environmental flows from other influences; and the Aral Sea monitoring program, established as part of the Syr Darya Control and Northern Aral Sea Project, will require trained staff in the newly

---

BOX 7.4

Monitoring Program for the Berg River Dam, South Africa

The record of decision for the Berg River Dam in South Africa required a detailed monitoring program to be established to provide the basis for an adaptive management framework implementing the ecological reserve. Thus the record of decision required sufficient baseline information to be collected prior to completion of the dam to assess the effectiveness of the environmental flows. The environmental flows will be revised if the monitoring demonstrates that the dam has an unacceptable ecological effect on the river or estuary.

The baseline monitoring program, initiated in 2002, included eight specialist studies for the riverine environment, nine specialist studies for the estuary, and a series of general catchment reports that included groundwater elements. The aim was to monitor the effects of the flow regime downstream of the dam. Data collection was completed in 2005, and a conceptual model was developed for determining and managing changes brought about by the dam. The program focused on the flow regime and the physical, chemical, and biological characteristics that the environmental flow was intended to support.

This comprehensive monitoring provides the baseline against which the project's environmental allocations are assessed and will be used to establish a comprehensive reserve for both the river and the estuary. The issue of appropriate flood releases is now under discussion in the light of advances in environmental flow assessment methodologies, the information available from the three-year baseline monitoring program, and concerns over water quality, especially salinity.

It is an example of best practice in environmental monitoring being used for adaptive management.

Source: Berg River Water Project (case study 11).

formed Kazakhstan Ministry of Natural Resources and Environmental Protection before it will be effective.

Integration

Although arguments for environmental flows usually rest on equity considerations, there can also be valuable economic benefits from the provision of environmental flows. Such a study was undertaken in the Senegal basin, where an analysis showed the high economic value of the floodplain. This had not been appreciated by senior decision makers in the Senegal basin countries and was instrumental in the agreement to permit flood releases from Manantali Dam. In the case of the Lesotho Highlands Water Project, the benefits of different flow scenarios to downstream communities, as well as the income lost from other uses of the water, were quantified in an economic study. Even though the downstream benefits were considerably less than the lost income, these economic arguments played a role in showing that there were real economic benefits from releasing environmental flows. The Berg River case study provides another example of the benefits of undertaking a full economic analysis of the value of water and the services that it supports.

EFAs are yet to be adopted as a mainstream part of EIA procedures for infrastructure assessment. The proposed Berg River Dam is the only project case study where the EFA was combined with the project EIA during the feasibility study. Although this is a sign that EFAs are being mainstreamed into project assessment, this integration remains a goal yet to be fully achieved in most projects. The Lesotho Highlands Water Project EFA was also a part of the project EIA, but was only completed after project appraisal because of the need to proceed rapidly to Phase 1B of the project in order to retain the workforce from Phase 1A and avoid startup costs. In hindsight, the goodwill of the Bank in agreeing to this accelerated commencement of Phase 1B had consequences for the subsequent decision-making process.

Cost-Effectiveness

The evidence, while quite limited, is that EFAs are often a relatively small fraction of the cost of new infrastructure developments. The costs of EFAs for new infrastructure projects can be divided into four components: (1) the cost of undertaking the EFA; (2) the cost of compensating affected downstream communities; (3) the cost of modifying the infrastructure; and (4) the cost of undertaking ongoing monitoring and enforcement.

Little information is available on any of these components of costs. The best information comes from the LHWP, where the comprehensive EFA was estimated to cost $2 million (0.07 percent of project costs) and the compensation was estimated to cost $14 million (0.5 percent of project costs). This included nearly two years of fieldwork to collect the baseline data and information and included the estimated costs of the resource losses to downstream communities. It included the potential impacts from both Phase 1 (now completed) and Phase 2 (yet to be commenced) of the project (see box 7.5).

An economic study that was part of the Berg River EFA showed that the cost of water released for downstream environmental purposes was likely to be substantial. With no environmental flows, it would be 5.9 years until an additional water supply scheme would be required for Cape Town. With "drought relief" environmental flow releases, this would be reduced to 4.9 years and would require an additional capital expenditure of about $6.6 million. With a "full maintenance" environmental flow rule, the time to a new water supply scheme would be reduced to 3.6 years, and the capital cost of providing the flows would be $14.9 million.

BOX 7.5
Economic Assessment of Downstream Impacts of the Lesotho Highlands Water Project

The findings of the environmental flow studies were subjected to a rigorous economic analysis, which concluded the following:

• There are sizable projected economic losses (in terms of use values and necessary compensation costs) to downstream communities ranging from M2.9 million (US$0.45 million) to M8 million (US$1.23 million) annually, depending on the environmental flow scenario chosen.

• Small increases in water releases from LHWP dams will have only a modest impact on these projected losses. Only considerable increases in environmental flows will succeed in sharply reducing the projected losses to downstream communities.

• These losses do not have a significant impact on the overall economic assessment of the project, as the projected losses are relatively small compared to the large benefits the project generates overall.

• From an economic point of view, the losses from reducing the yield of the project through higher environmental flows outweigh the benefits for downstream communities.

• The rate of return of the project is not seriously affected by changes in the environmental flow requirement scenario (see table 7.3). While the water transfer benefits of the project are reduced in the nontreaty scenarios and further reduce the already very low rate of return of the hydropower component, the rate of return of the overall project is only moderately reduced should one decide to increase releases, and the benefits to Lesotho and South Africa remain substantial. Thus it would be economically defensible to increase the environmental flows for ecological or social reasons, although doing so would not maximize the economic benefits of the project, and both parties would have to contend with reduced benefits of the project.

Source: Klasen 2002.

TABLE 7.3
The LHWP Economic Rate of Return for Different Flow Scenarios

Indicator	Treaty Scenario	Fourth Scenario	Design Limitation Scenario	Minimum Degradation Scenario
Annual economic value of losses ($ millions)	1.24	1.00	0.88	0.45
Lost variable royalty ($ millions)	0	6.36	33.82	71.85
% of total royalties	0	1.5	8.0	17.0
Economic rate of return (%)	7.6	7.4	7.3	7.1

Source: Watson forthcoming.

None of these studies of environmental flow provisions in new dams provided economic valuations of the benefits of the environmental flows to the downstream communities. However, other studies, such as the valuation of the ecosystem services from the flooding of the Hadejia-Nguru wetlands in northern Nigeria, have shown that, for highly productive downstream systems, these benefits can even be greater than the economic value of impounded water (Barbier, Adams, and Kimmage 1991).

The cost of reoperations of existing infrastructure, without infrastructure rehabilitation, can be relatively low. EFAs for reoperating existing infrastructure can be relatively inexpensive when carried out as a technical exercise without extensive stakeholder involvement and without modifications to infrastructure. For example, the Nature Conservancy was able to recommend detailed flow requirements to maintain key ecosystem processes to the U.S. Army Corps of Engineers (USACE) for inclusion in its comprehensive river basin plan for the Savannah River in the states of Georgia and South Carolina. This process took about nine months and cost US$75,000 (Nature Conservancy and Natural Heritage Institute forthcoming).

The Bridge River reoperation EFA, which was based on an extensive stakeholder involvement process, but did not require any infrastructure modifications, cost approximately $650,000, and the ongoing monitoring is estimated to cost about $550,000 per year. These costs, while substantial, are considerably less than the legal costs that BC Hydro would have faced if it had elected to maintain its stance that it was not responsible for uses of the water other than hydropower production, and it would not have benefited from the improved hydropower production that resulted from the new operating rules.

Retrofitting existing infrastructure to provide environmental flows can be very expensive. The LKEMP was essentially a project to help a heavily stressed ecosystem to recover from not undertaking and implementing an EFA when the dam was first built. While some of the cost of this project would have been incurred if the EFA had been undertaken originally, much of the $11 million can be attributed to the restoration effort, which resulted from inadequate environmental flow provisions. In another example, the outlet structures at both the Katse Dam (after the dam was constructed) and the Mohale Dam (when the dam was being designed) in the LHWP had to be modified at considerable cost to be able to provide the agreed environmental flows.

While no specific costs are available for the environmental component of the Tarim II project, it was clearly substantial given the need to reline hundreds of kilometers of canals with geomembranes and replace water control infrastructure. However, the financial benefits from the improved agricultural production alone reportedly were greater than the total cost of the project ($90 million) without including the economic benefits from the recovery of the lower Tarim River.

Influence

Environmental flows can lead to more efficient water use and benefit both environmental and consumptive water users. There were several win-win examples among the rehabilitation case studies. The reoperation rules for the Bridge River dams in Canada resulted in more power being generated as well as better delivery of water to downstream environments. Reoperation of infrastructure, particularly hydropower infrastructure, means a change in the timing, rather than the volume, of flows and so can result in these win-win situations. The Tarim basin upgrades, similarly, resulted in an increase in crop production as well as significantly increased flows to the downstream riverine environments. And the rehabilitation of the dams on the Syr Darya, together with improved operating procedures, meant that the environment of the Aral Sea as well as the fisheries industry have partially recovered, while more power has been generated from the hydropower dams and flooding has been reduced in upstream areas.

Environmental monitoring of environmental flows is essential for establishing baselines, undertaking enforcement, and implementing adaptive management. The monitoring program was established in the Berg River in 2002 in order to provide several years of baseline information against which the effects of the new dam can be measured. The underreporting of flow releases from the Lower Kihansi Dam was only detected because of an independent monitoring program conducted by the Rufiji Basin Water Office, and the quotas on water abstraction were only adhered to in the Tarim basin irrigation districts because of the monitoring and enforcement program.

Because the ecological response was uncertain in parts of the river downstream of dams in the Bridge River, the environmental flow plan required an adaptive management approach. The expected ecological response would be monitored under different flow release patterns, and the dam operating procedures would be modified after 2012 to employ the most effective release patterns. The monitoring program instituted as part of the environmental water releases in the LHWP shows that the river health targets have been met or exceeded except in two reaches, where there are problems with an endangered fish species (see table 7.4).

Successful environmental flow studies can have wider influence. Thus the recovery of Lake Sudoche has given the government of Uzbekistan the confidence to use environmental flows for the recovery of other degraded lakes; the partial recovery of the northern Aral Sea has led the government of Kazakhstan to consider reha-bilitating other affected water bodies; the procedures used to develop improved operations at the Bridge River hydropower dams have influenced other water-use plans in British Columbia; the experience of developing draft environmental flow rules for the Naraj Barrage on the Mahanadi River in India has led to the assign-ment of high priority for environmental flows in the new Orissa State water policy; and the DRIFT method, developed during the Lesotho highlands EFA, has now been applied in several other countries.

Institutional Drivers

Public concerns were a significant driver for rehabilitation and reoperation proj-ects. The restorations of the Aral Sea, Lake Sudoche, and the green corridor of the Tarim basin (see table 7.5) had the potential for providing such obvious ecosystem benefits to downstream people that the water needed for restoration was regarded as social and economic flows rather than environmental flows. In the case of the Senegal River basin, there were clear benefits to the mid-river and lower-river

TABLE 7.4
Lesotho Highlands River Condition Target Monitoring Results

Reach	Targeted River Condition Target	Measured River Condition Target	Measured Relative to Target
Reach 1	3	3	On target
Reach 2	4	3	Better
Reach 3	4	2–3	Better
Reach 4	3	2	Better
Reach 5	2	2	On target
Reach 6	2	3	Worse
Reach 7	4	3–4	Better
Reach 9	2	3	Worse

Source: Watson forthcoming.

TABLE 7.5
Drivers for New Infrastructure and Restoration Projects

Project	Judicial	Procedural	Evaluative	Instrumental	Professional	Public
New infrastructure						
Berg River		After 1998 the National Water Act provided a legislative driver for the conduct and implementation of the flow assessment			At an early stage, scientific groups advocated the inclusion of environmental flows in planning for the dam	Public awareness of the Cape Floral Kingdom acted as a backdrop to discussions about effects of the dam
Chilika lagoon		State water plan gave legitimacy to environmental flows but was not a specific driver		The World Bank required an EFA as part of the loan to reconstruct Naraj Barrage, but the recommendations are yet to be implemented		Pressure from lagoon communities to manage flooding and restore lagoon ecosystems was an indirect pressure to provide environmental flows
Lower Kihansi gorge			The Rufiji River Basin Office became an evaluator of implementation	Field-based review by government and World Bank provided basis for the restoration program		International NGOs place considerable pressure on the Tanzanian government and World Bank to maintain the threatened ecosystem

(continued)

TABLE 7.5
Drivers for New Infrastructure and Restoration Projects (*continued*)

Project	Judicial	Procedural	Evaluative	Instrumental	Professional	Public
LHWP				The World Bank safeguards initially did not include environmental flows in Phase 1A; subsequently, they were the major driver for Phase 1B	The South African experience and requirements acted as an indirect professional driver	
Rehabilitation and reoperation						
Aral Sea						Local communities were major drivers; extensive NGO publicity about degradation provided pressure for international action
Bridge River	Threat of legal action by federal department drove BC Hydro to develop an environmental flow investigation		An external review of power generation called for water user plans			Public concerns were expressed over the health of salmon stocks; environmental groups were advocating flow management

Senegal River	The Senegal water charter, when passed, contained clauses to provide downstream flows	The Word Bank became engaged late in project implementation when it provided funding for turbines	NGO studies showed the importance of restoring floodplains
Tarim basin	The National Water Law was supportive, but not a driver	The main driver was the government priority to protect downstream transport	

Source: Authors.

populations from maintaining some level of annual flood flows below Manantali Dam, and, although initially restricted in having their voices heard, these populations were assisted by NGOs to make their case. Although not a primary motivator, the rise in public concern about environmental degradation and threats to native fish was an important background driver leading to the reoperation of the dams on the Bridge River, Canada.

When the loss of ecosystem services is clearly apparent, there is no dispute about reestablishing environmental flows. The governments involved with two of the most effective restoration projects, the Aral Sea and the Tarim basin, had internalized environmental flows to the point where restoring flows to downstream areas was the major objective of the projects. This can be contrasted with the Mekong basin case study, where the term "environmental flows" was seen by some governments as a potential impediment to development.

Judicial drivers are seldom influential. The Bridge River case study provides the only example among the case studies where a judicial driver was one of the drivers for undertaking an EFA. In this case, the threat of being taken to court by a federal agency motivated BC Hydro to undertake a voluntary review of their operating rules to improve downstream environmental outcomes.

Procedural drivers support initiatives that are under way. In two of the cases— Chilika lagoon and Tarim basin—the state water plan and legislation were, respectively, consistent but not influential in initiating and implementing the EFAs. In other cases—Berg River, Kihansi gorge, and Senegal River—the procedural drivers provided supportive backing for EFA initiatives that were already under way.

NGOs can play a valuable role in bringing the need for environmental flows to public and government attention. NGOs contributed to the initiation of EFAs in several cases, but were a primary driving force only in the Lower Kihansi Power Project, where their pressure accelerated the response of both the Tanzanian government and the World Bank to restore the threatened ecosystem. The Rufiji Basin Water Office became arguably the most important institution for implementing the environmental flows. It was central to negotiating the agreed bypass flows as part of the final water right and to monitoring and enforcing the conditions of the water right.

When assessing large infrastructure projects, the need to address downstream issues, including the use of EFAs, needs to be more fully recognized and addressed in the planning and conduct of feasibility studies and EIAs. For new infrastructure, it is notable that in several cases the EIAs did not give adequate attention to downstream environmental flow issues when the projects were initially approved. This occurred with the Power VI Project (Kihansi gorge), where it was assumed that there were no sensitive downstream ecosystems; the LHWP, where it was assumed that a nominal minimum flow included in the transboundary agreement would meet downstream needs; and the Tarim Basin Project, where the Tarim I Project improved the management of the upstream irrigation districts but did not improve flows to rehabilitate the downstream green corridor. This was also the case in the initial development of the Manantali Dam on the Senegal River. In all cases, follow-up interventions were needed to rectify issues that emerged.

Summary of Project Lessons

The following are some of the lessons to emerge from the analysis of rehabilitation and reoperation projects:

- Major infrastructure investments can be required if large quantities of water are needed for the recovery of downstream environments; conversely, if the recovery of the downstream environments depends on changes in the timing rather than in the quantity of flows, then the investments can be relatively modest.
- The cost of EFAs for reoperation projects can be quite low, often less than $100,000.
- Reoperation and rehabilitation projects can sometimes provide win-win results for both downstream communities dependent on environmental flows and consumptive water users.

The following are some of the lessons to emerge from the analysis of new infrastructure projects:

- Catchment and basin water allocation plans provide benchmarks for reallocations of water when new infrastructure is being developed.
- There are real economic benefits when environmental flows are provided for downstream communities; economic studies that quantify these benefits can provide powerful arguments when water allocation decisions are being made.
- For new infrastructure, EFAs appear to be a small fraction of the cost of the development; retrofitting infrastructure because thorough EFAs were not undertaken when projects were being planned can be very expensive.
- Mechanisms to enforce environmental flow decisions, including monitoring programs, are essential: there can be strong pressures on infrastructure managers to limit environmental water releases.

Still, significant challenges remain for the integration of environmental flows into decisions about investment projects:

- Gaining acceptance that providing for environmental flows in development projects leads to increased social outcomes across all groups reliant on the water resource

- Introducing environmental flow concepts and methodologies for all activities that affect river flows and groundwater levels, including large-scale land use changes
- Assessing the impact of development activities on all downstream ecosystems, including groundwater-dependent ecosystems, estuaries, and coastal systems
- Building awareness among the environmental community so that EFAs become an integral part of the EIAs for project preparation and appraisal.

Note

1 Currently $10 billion has been allocated for returning water to the environment in the Murray-Darling basin in Australia.

PART IV

Mainstreaming Implications

CHAPTER 8

Achievements and Challenges

ENVIRONMENTAL FLOWS WORK within the World Bank is shaped by evolving global knowledge, practice, and implementation and helps to shape the repository of global knowledge and experience on environmental flows.

Providing water for the environment has now been institutionalized and mainstreamed in a growing number of developed countries such as Australia, New Zealand, the United States, and the countries of the EU. In these countries, the period of major water resources infrastructure development is now over (although the need to adapt to climate change may lead to a renewed interest in infrastructure investment), and the focus has been on the rollout of basin- or catchment-level water allocation plans that include environmental water provisions. South Africa, a country with an economy in transition, is also preparing to undertake nationwide catchment water resources plans. In all these countries, there is both broad acceptance of the importance of protecting the aquatic environment and general support for environmental water provisions in the water allocation plans.

In these countries, the focus of environmental flow assessments has continued to broaden beyond the provision of water to rivers and associated wetlands to include other hydrologic components, including estuaries, near-shore areas, and linked groundwater systems. However, there is not yet the same level of expertise and depth of experience in assessing the environmental water requirements for these new areas as there is for downstream river systems.

In developing countries, in contrast, there has been a greater focus on the assessment of the downstream impacts of new infrastructure or the restoration of downstream aquatic ecosystems that have been degraded from existing infra-structure.[1] While the World Bank has not had much experience in major basin water allocation planning programs, it has been a partner in some of the more notable achievements in ecosystem restoration in developing countries, such as China (Tarim basin), the Senegal basin, and Central Asia (northern Aral Sea).

The science of environmental flows has advanced considerably in the last 20 years from a focus on individual aquatic species (although there are circum-stances where this continues to be relevant) to a much broader concern about ecosystem protection or restoration. Many scientifically credible, field-tested assess-ment methods are now available, ranging from simple desktop approaches to complex, field-based holistic methods. Some of the holistic methods combine hydrologic and environmental science with social and economic assessments. There is now also experience in organizing multidisciplinary teams to carry out these holistic assessments.

International development organizations and NGOs have been active in promoting understanding of environmental flows in developing countries, holding training courses, providing assistance in conducting EFAs, and devel-oping support materials and information sources. The World Bank has contributed to this richer information environment through its BNWPP envi-ronmental flows window as well as the development of support documentation, which has been distributed widely.

Scientific Achievements

There have been considerable advances in the science of environmental flows over the last 15 years, including improvements in basic scientific understanding and the development of EFA techniques.

Understanding Environmental Flows

Hydrological knowledge and ecological knowledge have advanced considerably, so that there is a now much better understanding of the dependence of both species and ecosystems on flows in freshwater systems as well as a broadly agreed conceptualization among scientists about the way to define the parameters for riverine flows in an ecologically meaningful way.

The ecological response of wetlands and floodplains to different flow regimes is quite well known in many parts of the world. There is greater understanding of the flow needs of in-stream species (especially fish and invertebrates). The effect of disturbances on the food web is increasingly well understood, and the effects of different flow regimes on substrate and physical habitat are improving. However, these and other advances in knowledge are usually limited to the local regions where the scientific information was collected; it remains difficult to develop useful generalizations that can be applied more widely.

However, the same level of understanding of ecological responses is not yet available in the new areas where EFAs are being applied such as linked groundwater systems, estuaries, and near-shore areas.

Hydrological science has advanced to the point where a wide range of river system models are available (although the accuracy of these model predictions is almost always limited by the absence of good flow data) that can provide ecologically useful flow predictions. Hydraulic models can estimate the level and velocity of flows, and the first generation of floodplain hydraulic models is available that can predict the extent (and sometimes duration) of floodplain wetting.

EFAs require the integration of information from a range of scientific disciplines: hydrology, ecology, geomorphology, and hydrogeology. In some cases, this integration includes information from the economic and social sciences. There has been increasing experience in forming teams of scientists and experts from these different disciplines who can work together in spite of different terminologies, approaches, and scientific cultures.

Development of Environmental Flow Assessment Methods

Over the last 15 years, a considerable body of experience has developed in applications of the extensive range of EFA techniques to evaluating the impacts of individual projects as well as undertaking basinwide studies. This experience is available through publications and a Web site.[2] Much support material has been produced for environmental flow applications, including technical documents, a newsletter, Web pages, and a periodic conference on environmental flows methods and applications. There is sufficient understanding of their strengths and weaknesses to be able to customize the techniques to suit each application. This customization is illustrated in the Mekong basin and Pangani case studies (case studies 7 and 8).

Current Challenges

First, the current terminology gives rise to confusion. The misperceptions that arise from the term "environmental flows" can lead to a rejection of environmental flow assessments by managers. While it would be helpful to adopt a new term, such as "social flows" or "environmental and social flows," the reality is that "environmental flows" is so widely used that it would be very difficult to get acceptance for a new term. Retention of this terminology means that there is a need to stress, whenever the term is used, that environmental flows are intended to provide healthy river systems and that these bring benefits to many groups in society.

Second, there is a need to incorporate impacts on water other than rivers. EFA techniques have been developed primarily to assess the effects of changes in river flows on ecosystem services. The techniques need to be extended to include impacts on lakes, groundwater, estuaries, and near-shore systems (Young 2004). While these water bodies have been included in some EFA assessments, including in some of the present case studies, there is not yet a systematic procedure for integrating these nonriverine components of the hydrologic cycle into EFAs.

Third, the impacts of land-use changes and land management activities on river flows and groundwater systems is still not properly integrated into EFAs. Some useful empirical relationships can be used to estimate the annual average interceptions of water by land-use changes, but there is still insufficient information available to make detailed assessments of the downstream environmental impacts of these activities (Zhang, Dawes, and Walker 1999).

Fourth, the ecosystem services that people rely on will be affected in complex ways by the changes in the volume and timing of flows induced by climate change that are not yet properly understood (see box 8.1). In addition, climate change will have an influence on the demand for water for irrigation, industry, and municipalities. These shifts in location, quantity, timing, and sources of water demand will have implications for providing water for environmental services.

BOX 8.1
Climate Change and Evapotranspiration

The natural assumption is that, with the observed increase in air temperatures attributed to climate change, there would be a corresponding increase in the potential rate of evaporation. However, the evidence is that the atmospheric demand, as measured by pan evaporation, has been decreasing over the past 50 years.

The reason for this counterintuitive result lies in the fact that evaporation is more sensitive to changes in net radiation, vapor pressure deficit of the air, and wind speed than to air temperature. Because vapor pressure has increased with global temperatures, relative humidity has remained about the same. Consequently, pan evaporation is particularly sensitive to wind speed. Average wind speed reportedly has declined in Australia, China, India, New Zealand, Thailand, the Tibetan Plateau, and the United States. This has been the main driver of the observed decreases in pan evaporation. It is difficult to assess whether these reductions in long-term average wind speed are local effects attributable to changes in the immediate environment of the pans (for example, growing trees or other obstacles progressively obstructing the air flow) or a more regional phenomenon.

Source: Roderick and others 2007.

Climate change will act as a catalyst for choice: environmental assets and ecosystem services will need to be reassessed, with the essential assets and ecosystem services being identified and protected. The multiple effects of climate change on environmental flows have yet to be factored into EFAs and water allocation plans in a systematic way.

Fourth, where EFAs have been carried out for surface water and groundwater resources, the assessments have usually been carried out separately. However, in many cases hydrological interdependencies exist between surface water and groundwater, and the assessments should be undertaken for the integrated system. Thus some environmental assets depend on both surface water and groundwater at different times of the year. The maintenance of the ecosystem services from these assets will require the joint planning of surface water and groundwater availability. However, with the current lack of understanding of physical connectivity and, in many cases, social dependence on the joint resource, the assessment of environmental flow needs is usually undertaken separately for surface water and groundwater systems.

Fifth, in general, the EIA community has yet to integrate the assessment of environmental flows into EIAs (for project-level assessments) and SEAs (for more strategic-level assessments). Partly this is because EIAs and SEAs were developed by the environment sector, while EFAs were developed within the water resources sector. This represents a major disconnect between the water and environment communities and is counterproductive for promoting environmentally responsible development. For the Bank, environmental flows need to be fully integrated into the planning, design, and operations of infrastructure projects and into environmental assessments.

Another aspect of this integration at a higher level of decision making is to ensure that water resources and environmental policies and laws are harmonized within countries. The Bank can contribute to this outcome by promoting harmonization when assisting with policy reforms.

Finally, environmental flows are based on the concept of water sharing—that is, the idea that flows in river or groundwater systems should be shared equitably. Upstream infrastructure projects typically generate considerable economic benefits, and, in many cases, these benefits accrue to populations that are distant from the water sources.

Benefit sharing provides an alternative approach to water sharing, where the economic benefits from the development project are shared with the affected people both upstream and downstream of the development. Environmental flows are concerned with delivering goods and services by enhancing and conserving the environment; benefit sharing in this context is first about recognizing the "voiceless" downstream communities and bringing them into the decision making and second about both sharing the benefits of development as well as fairly compensating people for the loss of environmental goods and services that are linked to a reduced or altered flow regime.

Integrating Environmental Flows into Decisions

Environmental flow considerations were initially introduced into assessments for new infrastructure projects. They are now being gradually mainstreamed into more strategic levels of decision making, including national water resources policies and the formulation of basin- and catchment-level water plans.

Policy Achievements

Environmental flow considerations have been incorporated into the water resources policies of a number of countries, as illustrated by the case studies of Australia, the EU, Florida, South Africa, and Tanzania in this document. While most of the examples of environmental flow recognition in policies so far have been in economically developed countries, an increasing number of developing countries are considering the inclusion of environmental flows.

Good progress has been made in implementing the policy provisions for environmental flows under the Floridian, Australian, and South African policies. In Florida, minimum flows and levels have now been established in 237 water bodies and additional minimum flow levels continue to be established, although not to the originally proposed schedule. In Australia, environmental flow provisions have been incorporated into 120 catchment and groundwater allocation plans, although, as in Florida, these have slipped behind the agreed schedule. South Africa has established interim ecological reserves in all catchments.

The World Bank has played an important role in assisting some developing countries and regions to introduce environmental flow considerations into their water policies as shown in Tanzania's national water policy, Orissa State's water policy, and the Senegal basin water charter described in the case studies. The Bank is currently providing technical assistance to Mexico in the revision of its water policy and has identified assistance with policy reform in the China CWRAS. It is also supporting policy dialogue on the energy sector in some Indian states and on water resources in general in Pakistan.

Plan Achievements

Catchment plans with environmental water provisions have been rolled out across Europe under the Water Framework Directive and across Australia under the water reform agenda, while South Africa and Tanzania are preparing to develop basin-level water resources plans with environmental flow provisions. Although water resources plans are yet to be established in the catchments of the Kruger

National Park in South Africa, this region is particularly noteworthy because of its influence both nationally and internationally in the development of approaches to the inclusion of environmental flow considerations in catchment planning.

The Bank has contributed by assisting with basin-level water resources planning, including environmental water allocations in the Mekong basin and the Senegal basin. In the Mekong basin, an international environmental flows expert was introduced to the Mekong River Commission through BNWPP technical assistance, and the World Bank was the implementing agency for the GEF Water Utilization Project, which provided assistance for implementation of the 1995 Mekong agreement with its environmental flow provisions. In the Senegal basin, the Bank provided support for implementation of the Senegal basin water charter and provision of environmental water releases following the installation of turbines at Manantali Dam.

Infrastructure Projects

There are a growing number of examples where environmental flows have been incorporated into both the operations of new infrastructure and the rehabilitation and reoperation of existing infrastructure.

The revision of the operating rules for the dams on the Bridge River in Canada resulted in both improved environmental outcomes and increased power production. This project was also noteworthy for giving responsibility to the stakeholder committee to develop the reoperation rules for the dams and for crafting an adaptive management plan that included alternative release schedules to determine which schedule provided the greatest downstream benefits. In another example, the Nature Conservancy has established a successful partnership with the U.S. Army Corps of Engineers and is reviewing the operating rules at 26 dams in the United States. The Nature Conservancy provides technical guidance on the operations of the various USACE dams.

The EFA carried out for the Berg River Dam in South Africa provides a good-practice example of EFAs for new infrastructure development. It was the first large water resources infrastructure development project in South Africa to be designed, constructed, and operated within the framework of the National Water Act, with provisions for basic human needs and the ecological reserve. It was developed in accordance with the guidelines of the World Commission on Dams. It also illustrates an adaptive approach to the determination of environmental flow requirements, including a preliminary assessment carried out as part of prefeasibility investigations, a more detailed assessment as part of the feasibility studies, and a series of subsequent workshops and specialist meetings to obtain detailed inputs from specialists. A comprehensive determination of the ecological reserve is expected following finalization of the three-year monitoring program.

The World Bank has been a significant contributor to the growing body of experience in this area, particularly through its ability to extend its support through all stages of project decision making, from initial discussion, through conduct of EFAs, to implementation of the agreed flow regimes, to support for monitoring and enforcement programs. The Bank has also used its convening powers to facilitate discussions and support scientific studies and dialogue to reach agreements among nations over environmental flows, especially in select transboundary cases. Between 2002 and 2004, the governments of the Senegal basin signed the water charter, which included provisions for flows to maintain important downstream ecosystem functions. The operating rules for the Manantali Dam now include water releases to provide artificial floods for parts of the mid-river floodplains. In addition, programs were implemented to release water through the river embankments to reinundate the Diawling National Park in the delta using water stored behind the Diama Dam near the mouth of the river.

The Bank-supported Lesotho Highlands Water Project has established an environmental flow policies and procedures for operating the Katse and Mohale Dams. This constitutes the first systematic effort by the World Bank to support the development and implementation of downstream mitigation and compensation programs during project development. It has been described by an independent audit as being "at the forefront of global practice" (Lesotho Highlands Development Authority 2007; see box 8.2).

BOX 8.2
Achievements of the Lesotho Highlands Water Project

The DRIFT method for environmental flow assessment was developed and applied during this project by leading South African environmental flow consultants. It makes an important contribution to the science of environmental flows. It is the first fully integrative methodology combining environmental, social, and economic factors in assessing the impacts of different flow scenarios. It has now been applied to other environmental flow studies within South Africa and, in modified form, in the Pangani and Mekong basin studies.

The EFA was carried out thoroughly enough to convince skeptical development authorities of the case for providing for downstream water-dependent communities, including economic studies into the financial effects of providing different levels of environmental flow releases.

The thorough social surveys found that approximately 39,000 people directly and indirectly dependent on water and water-related resources would be affected downstream of the dam—many times the original estimate and an order of magnitude higher than the number of people affected upstream of the dam.

The Bank's support for the recovery of the northern Aral Sea and the Tarim Basin II Project has also yielded significant outcomes. The Aral Sea had been widely publicized as an example of an ecosystem that was virtually unrecoverable. However, the rapid recovery of the northern Aral Sea under the World Bank's Syr Darya Control and Northern Aral Sea Project has shown what can be achieved with adequate funding, strong local and government commitment, strong Bank leadership, and an understanding of technical and operational impediments at upstream infrastructure.

The Tarim River was a severely degraded ecosystem that was imposing considerable costs on local communities and posing a strategic threat to one of China's major transport routes. With World Bank technical and financial assistance, the Chinese provincial government restored flows to the lower river and Lake Taitema, while also increasing agricultural production and incomes in the irrigation districts.

Resource Material and Assistance

International development organizations and NGOs have produced many support documents, Web sites, databases, training courses, and discussion for countries interested in undertaking EFAs for project proposals or for basin plans. The World Bank has contributed to these support materials through its water resources and environment technical notes on environmental flows.

As a result of the EFA studies and economic analysis, the original minimum flows stipulated under the 1986 Lesotho Highlands Water Project treaty were increased by a factor of 3 and 4 for the Mohale and Katse Dams, respectively. The Mohale Dam outlet valves were resized to accommodate the anticipated higher flows, and a new valve was added to Katse Dam to accommodate higher EFA releases. Compensation payments have been negotiated for the remaining losses in ecosystem services for downstream communities, using both a negotiated formula involving distance from the dam and the results of the monitoring program.

A monitoring program has been established, and early indications are that, under the agreed flow release policy, the river health targets have been met or exceeded in all except two reaches.

The project outcomes included better than predicted ecological impacts and compensation to downstream communities, with little impact on the project's economic rate of return. This best-practice work has contributed to improving the political image of a high-risk project that has faced two inspection panel complaints and major corruption charges.

Source: Watson forthcoming.

The BNWPP-funded environmental flows expert panel has provided assistance to 16 countries through a mixture of training courses, workshops, and assistance in introducing and undertaking EFAs. A number of international development organizations and NGOs have run training courses in environmental flows that have been influential, such as the courses run by IUCN in Mesoamerica to develop a network of informed and influential champions for environmental flows.

Notes

1 There are exceptions such as Tanzania, which is embarking on a basin water allocation planning program, including environmental flow provisions.
2 See Postel and Richter (2003); http://dw.iwmi.org/ehdb/wetland/index.asp.

C H A P T E R   9

Framework for Mainstreaming Environmental Flows

ENVIRONMENTAL FLOWS ARE CENTRAL TO SUPPORTING sustainable development, sharing benefits, and addressing poverty. In some circumstances, environmental flow assessments can also lead to more efficient water use and benefits to both environmental and consumptive water users. Effective integration of environmental flows in decision making is a necessary requirement for promoting environmentally responsible water resources development in the face of changing societal values and reduced availability of water under climate change. It is also critical to promoting environmentally responsible climate change adaptation strategies. It needs to be an integral part of programs for sharing benefits from water infrastructure development. In order to achieve environmentally sustainable and socially responsible development, more systematic and timely attention will need to be paid to downstream impacts using scientifically credible EFA methods as countries, through both public and private sector investments, expand their infrastructure, especially dams, in many sectors.

The Way Forward

The overall goal of this report is to advance the Bank's understanding and integration in operational terms of environmental water allocation into integrated water resources management. Achieving this goal is essential for supporting the implementation of several recent Bank strategies and action plans—including the infrastructure action plan, investments in hydropower, the Agriculture Water Management Initiative, the water supply and sanitation business plan, and the Strategic Framework for Climate Change and Development—in an environmentally responsible manner, consistent with the SDN vision to mainstream environment in World Bank operations.

The preceding chapters demonstrate how central environmental flows are to IWRM, how they will be affected by the impacts of climate change, and how they are central to climate change adaptation responses in the water sector. They highlight the evolving understanding of, and knowledge about, environmental flows as well as the integration of environmental flows into water resources decision making at the policy, plan, and project levels within and outside the Bank. They also illustrate the complexities and challenges associated with the implementation of environmental flows across the many sectoral uses of water, spanning many parts of the world.

There is a growing body of experience in implementing environmental flows, including monitoring and adaptation of management procedures. Chapters 5, 6, and 7 summarize lessons from the integration of environmental flows in the formulation and implementation of water policies, river basin or catchment plans, and design and operations of infrastructure development and rehabilitation projects.

The earlier chapters provide indicators of the direction that the Bank and its clients need to take to support better integration of environmental flows in policy reforms, river basin plans, land-use change and watershed management projects, and infrastructure investment planning, design, and operations. *A key lesson is the high financial, social, reputational, and political costs associated with not undertaking (or undertaking late) a thorough EFA when projects are being prepared.* The evidence, while limited, is that the cost of conducting an EFA for new infrastructure developments often is relatively small, whereas the cost of retrofitting existing infrastructure to increase the capacity and provide the flexibility for environmental flows can be very high. However, if environmental flow releases are obtained through the reoperation of an existing hydropower dam with minimal stakeholder involvement and no modifications to the infrastructure, then the costs can be relatively low—on the order of $50,000–$75,000 based on the Nature Conservancy's experience with the U.S. Army Corps of Engineers.

Another important lesson concerns the critical links between environmental flows and riverine community livelihoods that are underscored by two African case studies (the Senegal basin and the Lesotho Highlands Water Project). In the former, the water charter signed by the governments of Mali, Mauritania, Senegal, and Guinea recognized the provision of flows to water the mid-river floodplain and ensured the maintenance of agricultural and fishing activities.

The LHWP broke important new ground not only in supporting the development and application of a state-of-the-art EFA methodology, but also in applying a well-structured approach that links resource losses associated with reduced river flows to community livelihoods and addresses social impacts related to environmental flows. In the absence of any known clearly defined methods, procedures, and guidelines globally for addressing downstream social impacts of dams, the LHWP environmental flow experience offers important lessons in the following areas:

- Understanding the difference between downstream social impacts and upstream social impacts
- Recognizing the difference in magnitude in the number of people who can be affected downstream of the dam (about 39,000) compared to upstream of the dam (around 4,000)
- Developing an approach for systematically defining the affected communities (or "the population at risk") downstream of dams
- Delineating the downstream socioeconomic impacts associated with changes in river flows
- Defining approaches for addressing and mitigating the social impacts associated with significant changes in river flows and their limitations (in addressing impacts in proximal reaches versus distal reaches)
- The challenges of developing and implementing a successful environmental flow policy for operating new dams.

The successful EFAs in Bank project implementation was a result of strong leadership by individual task team leaders and other factors, rather than formal requirements or formal directions for initiatives related to integrating environmental flows or restoring degraded ecosystems. However, mainstreaming EFA in water resources decision making (under a better business model) will require a fundamental shift by the Bank from an ad hoc approach to an institutionalized approach—that is, more structured, systematic, and timely—to support the integration of EFA into Bank water resources infrastructure planning, design, and operations and policy dialogue.

There are also clear lessons from the directions taken by countries such as Australia, South Africa, and the countries of the EU that the Bank can learn from as it moves forward to support implementation of the SDN vision:

- Environmental flow considerations need to be moved up to the more strategic levels of policy and basin plans to ensure that there is a strong basis for environmental water allocations.
- Where possible, environmental flow considerations should be integrated with existing IWRM and environmental assessment processes, such as the development of basin-level water resources plans.

- Environmental flows should be concerned with any development activities, including land use changes, that affect flow regimes with consequences for downstream ecosystems and dependent communities.
- The full range of downstream aquatic ecosystems, including estuaries and near-shore areas, should be included in EFAs.
- Environmental flow assessments should integrate surface water and groundwater requirements where there are linked systems.
- Climate change effects—through changes in water availability, changes in environmental assets, and changes in demand patterns—will be an important consideration in including environmental flows in policies, making provisions in basin or catchment plans, and assessing project-level impacts.

Finally, several international development organizations, NGOs, and research organizations have accumulated considerable expertise in undertaking EFAs and providing training and other support to developing countries. Their experience can complement the Bank's ability to convene development partners and work with developing countries throughout the full decision-making process.

A Framework for Bank Action

The proposed framework for action consists of four components: strengthen Bank capacity, strengthen environmental flow assessments in project lending, promote the integration of environmental flows at the policy and planning levels, and expand collaborative relationships. These actions are summarized in table 9.1.

The following guidelines are intended to strengthen Bank capacity:
- Promote the development of a common understanding across the water and environmental communities about the concepts, methods, and good practices related to environmental flows, including the need to incorporate EFAs into environmental assessment at both project (EIAs) and strategic (SEAs) levels.
- Build the Bank's in-house capacity in EFA by broadening the pool of ecologists, social scientists, and environmental and water specialists trained in EFA.

The following guidelines are intended to strengthen environmental flow assessments in project lending:
- Disseminate existing guidance material (from within and outside the Bank) concerning the use of EFAs in program and project settings and conduct training for Bank and borrower-country staff on this emerging issue.
- Identify settings, approaches, and methods for the select application of EFAs in the preparation and implementation of project-level feasibility studies and as part of the planning and supervisory process.
- Provide support for hydrological monitoring networks and hydrological modeling to provide the basic information for undertaking EFAs.

TABLE 9.1
A Framework for Adopting and Integrating Environmental Flows into Bank Work

Outcome	Decision Level	Bank Instrument	Support Material	Collaboration
Integrate EFA into planning studies for infrastructure, including EIA and SEA	New investment programs and projects; rehabilitation or reoperation projects	Increase focus on downstream issues in program and project design, making use of existing technical notes on EFA	Environmental assessment update on environmental flows; training materials; other support materials, including case studies	Collaborate with experienced international agencies and NGOs, the International Hydropower Association, and other relevant industry groups
Integrate downstream social impacts in infrastructure planning	New investment programs and projects; rehabilitation or reoperation projects	Increase focus on downstream social issues in program and project design making use of Bank experiences in previous projects (for example, LHWP)	Technical note on downstream social issues, impacts, and mitigation and compensation options; training materials; other support materials including case studies	Collaborate with SDV on the initiative for enhancing local benefits from hydropower projects
Broaden application of EFA to noninfrastructure projects	New investment projects	Include environmental flow consideration in CAS and CWAS; test application of EFA to select operations of this type	Environmental assessment update on environmental flows; training materials; other support materials	Draw on experiences in countries where interception activities are assessed for their flow impacts
Ensure that EFAs include all affected downstream ecosystems	Investment and non-investment projects and basin and catchment plans	Increase focus on downstream issues in program and project design, making use of existing technical notes on EFAs	Technical documents; training materials	Collaborate with experienced international agencies and NGOs

(continued)

133

TABLE 9.1
A Framework for Adopting and Integrating Environmental Flows into Bank Work (*continued*)

Outcome	Decision Level	Bank Instrument	Support Material	Collaboration
Promote inclusion of EFAs into basin and catchment plans	Basin and catchment plans	Include EFAs in proposals for land use plans in CAS and CWRAS	Technical documents; training materials	Collaborate with experienced international agencies and NGOs
Promote inclusion of environmental flow considerations in water resources and environment policies	National policy and transboundary agreements	Include downstream ecosystem impacts in proposals for environment and water resources policies in CAS and CWAS	Technical and analytical documents; training materials	Draw on experiences in countries that have implemented water policies with environmental flow components
Harmonize sectoral policies with water resources policy	National policy	Include harmonization with water resources and environment policies in proposals for policy reform in CAS	Technical and analytical documents	Draw on experiences in countries that have implemented water and other sectoral policies with environmental flow components

Source: Authors.

- Prepare an update for the *Environmental Assessment Sourcebook* concerning the use of EFAs in SEAs and EIAs.
- Prepare a technical note that defines a methodology for addressing downstream social impacts of water resources infrastructure projects.
- Test the application of EFAs to include infrastructure other than dams (such as levees and dikes for flood protection and excessive groundwater pumping) that can affect river flows as well as noninfrastructure activities, such as investments in large-scale land-use change and watershed management, that affect downstream flows and ecosystem services.
- Broaden the concept of environmental flows for appropriate pilot projects to include all affected downstream ecosystems, including groundwater systems, lakes, estuaries, and coastal regions.
- Develop support material for Bank staff and counterparts in borrowing countries, such as case studies, training material, technical notes, and analyses of effectiveness.

The following are intended to promote the integration of environmental flows at the policy and planning levels:

- Promote basin plans that include environmental flow allocations, where relevant, through country dialogue.
- Use CASs and CWRASs to promote Bank assistance with basin or catchment planning and water policy reform so that the benefits of environmental water allocations for poverty alleviation and the achievement of the MDG are integrated into country assistance.
- Incorporate environmental water needs into Bank SEAs such as country environmental assessments and sectoral environmental assessments.
- Test the use of EFAs in a small sample of sectoral adjustment lending operations, including where the sectoral changes will lead to large-scale land-use conversion.
- Promote the harmonization of sectoral policies with environmental flow concepts in developing countries and the understanding of sectoral institutions about the importance of considering the impact of their policies on downstream communities.
- Develop support material for Bank staff on the inclusion of environmental flows into basin and catchment planning and into water resources policy and legislative reforms.
- Draw lessons from developed countries that have experience in environmental flows in catchment planning.

Finally, expand collaborative relationships through the following:

- Expand collaboration with NGOs (IUCN, the Natural Heritage Institute, The Nature Conservancy, WWF, and others), international organizations (IWMI, Ramsar Secretariat, UNEP, and UNESCO), and research institutions to take advantage of their experience in conducting EFAs and building environmental flows capacity in developing countries.

- Strengthen collaborative relationships with industry associations, such as the International Hydropower Association, and private sector financing to extend their current recognition of environmental flows as desirable hydrological outcomes to include the social and economic outcomes that result from the ecosystem services delivered by downstream flows.
- Integrate the lessons from the economic and sector work into—and coordinate the activities outlined above with—the ongoing initiative of the Bank's Sustainable Development Network and the Energy, Transport, and Water Department for enhancing the benefits to local communities from hydropower projects.

By adopting this framework, the Bank will be better placed to fulfill its strategy of encouraging more investment in water resources infrastructure using a new business model that includes better consideration of the environmental, social, and economic impacts of the investment.

PART V

Appendixes

The Brisbane Declaration

ENVIRONMENTAL FLOWS ARE ESSENTIAL *for freshwater ecosystem health and human well-being.*[1]

This declaration presents summary findings and a global action agenda that address the urgent need to protect rivers globally, as proclaimed at the tenth International River Symposium and International Environmental Flows Conference, held in Brisbane, Australia, on September 3–6, 2007. The conference was attended by more than 750 scientists, economists, engineers, resource managers, and policy makers from more than 50 countries.

Key findings include the following:

Freshwater ecosystems are the foundation of our social, cultural, and economic well-being. Healthy freshwater ecosystems—rivers, lakes, floodplains, wetlands, and estuaries—provide clean water, food, fiber, energy, and many other benefits that support economies and livelihoods around the world. They are essential to human health and well-being.

Freshwater ecosystems are seriously impaired and continue to degrade at alarming rates. Aquatic species are declining more rapidly than terrestrial and marine species. As freshwater ecosystems degrade, human communities lose important social, cultural, and economic benefits; estuaries lose productivity; invasive plants and animals flourish; and the natural resilience of rivers, lakes, wetlands, and estuaries weakens. The severe cumulative impact is global in scope.

Water flowing to the sea is not wasted. Freshwater that flows into the ocean nourishes estuaries, which provide abundant food supplies, buffer infrastructure against storms and tidal surges, and dilute and evacuate pollutants.

Flow alteration imperils freshwater and estuarine ecosystems. These ecosystems have evolved with, and depend upon, naturally variable flows of high-quality fresh water. Greater attention to environmental flow needs must be exercised when attempting to manage floods; supply water to cities, farms, and industries; generate power; and facilitate navigation, recreation, and drainage.

Environmental flow management provides the water flows needed to sustain freshwater and estuarine ecosystems in coexistence with agriculture, industry, and cities. The goal of environmental flow management is to restore and maintain the socially valued benefits of healthy, resilient freshwater ecosystems through participatory decision making informed by sound science. Groundwater and floodplain management is integral to environmental flow management.

Climate change intensifies the urgency. Sound environmental flow management hedges against potentially serious and irreversible damage to freshwater ecosystems from climate change impacts by maintaining and enhancing ecosystem resiliency.

Progress has been made, but much more attention is needed. Several governments have instituted innovative water policies that explicitly recognize environmental flow needs. Environmental flow needs are increasingly being considered in water infrastructure development and are being maintained or restored through releases of water from dams, limitations on groundwater and surface water diversions, and management of land use practices. Even so, the progress made to date falls far short of the global effort needed to sustain healthy freshwater ecosystems and the economies, livelihoods, and human well-being that depend upon them.

Global Action Agenda

The delegates to the Tenth International River Symposium and Environmental Flows Conference call upon all governments, development banks, donors, river basin organizations, water and energy associations, multilateral and bilateral institutions, community-based organizations, research institutions, and the private sector across the globe to commit to the following actions for restoring and maintaining environmental flows:

Estimate environmental flow needs everywhere immediately. Environmental flow needs are currently unknown for the vast majority of freshwater and estuarine ecosystems. Scientifically credible methodologies quantify the variable—not just minimum—flows needed for each water body by explicitly linking environmental flows to specific ecological functions and social values. Recent advances enable rapid, regionwide, scientifically credible environmental flow assessments.

Integrate environmental flow management into every aspect of land and water management. Environmental flow assessment and management should be a basic requirement of integrated water resource management; environmental impact assessment; strategic environmental assessment; infrastructure and industrial development and certification; and land-use, water-use, and energy production strategies.

Establish institutional frameworks. Consistent integration of environmental flows into land and water management requires laws, regulations, policies, and programs that (1) recognize environmental flows as integral to sustainable water management, (2) establish precautionary limits on allowable depletions and alterations of natural flow, (3) treat groundwater and surface water as a single hydrologic resource, and (4) maintain environmental flows across political boundaries.

Integrate water quality management. Minimizing and treating wastewater reduce the need to maintain unnaturally high stream flow for dilution purposes. Properly treated wastewater discharges can be an important source of water for meeting environmental flow needs.

Actively engage all stakeholders. Effective environmental flow management involves all potentially affected parties and relevant stakeholders and considers the full range of human needs and values tied to freshwater ecosystems. Stakeholders suffering losses of ecosystem service benefits should be identified and properly compensated in development schemes.

Implement and enforce environmental flow standards. Expressly limit the depletion and alteration of natural water flows according to physical and legal availability and accounting for environmental flow needs. Where these needs are uncertain, apply the precautionary principle and base flow standards on best-available knowledge. Where flows are already highly altered, utilize management strategies, including water trading, conservation, floodplain restoration, and dam reoperation, to restore environmental flows to appropriate levels.

Identify and conserve a global network of free-flowing rivers. Dams and dry reaches of rivers prevent fish migration and sediment transport, physically limiting the benefits of environmental flows. Protecting high-value river systems from development ensures that environmental flows and hydrological connectivity are maintained from river headwaters to mouths. It is far less costly and more effective to protect ecosystems from degradation than to restore them.

Build capacity. Train experts to scientifically assess environmental flow needs. Empower local communities to participate effectively in water management and policy making. Improve engineering expertise to incorporate environmental flow management in sustainable water supply, flood management, and hydropower generation.

Learn by doing. Routinely monitor relationships between flow alteration and ecological response before and during environmental flow management and refine flow provisions accordingly. Present results to all stakeholders and to the global community of environmental flow practitioners.

Notes

1 *Environmental flows* describe the quantity, timing, and quality of water flows required to sustain freshwater and estuarine ecosystems and the human livelihoods and well-being that depend on these ecosystems.

APPENDIX B

Infrastructure Design
Features for Environmental
Flows from Dams

THE FOLLOWING DESCRIPTION of the physical infrastructure features that are needed to deliver environmental flows is taken from a recent report to the World Bank (Nature Conservancy and Natural Heritage Institute forthcoming).

Water Release Infrastructure

Water release infrastructure, including variable outlet and turbine-generator capacities as well as multilevel (selective withdrawal) outlet structures, can affect the capacity of a dam to release environmental flows. This section deals with these in turn.

Variable Outlet and Turbine-Generator Capacities

The ability of a dam operator to provide a range of flows for downstream environmental purposes is ultimately dependent on a dam's outlet and turbine-generator capacities. Many hydropower dams lack adequate turbine-generator capacity to make large releases—such as the controlled floods that may be highly desirable for maintaining the ecological health of downstream floodplain ecosystems and estuaries—without sacrificing power generation. Because of these constraints, some fraction of controlled flood discharges must be released through the dam's flood spillway. This sacrifice of power generation causes dam operators to resist such controlled flood releases. This is the situation at the Manantali Dam in the Senegal River basin. At that dam, some 2,000 cubic meters per second of water would need to be released to inundate the floodplain to support 50,000 hectares of recessional agricultural production, yet the outlet and turbine-generator capacity is only capable of delivering 480 cubic meters per second. The rest of the required flow would need to be released through the spillway, thereby compromising hydropower generation. Necessary structural modifications to expand the power-house capacity from 480 to 2,000 cubic meters per second would be very expensive at this point. Had the powerhouse capacity and reservoir storage tradeoff been opti-mized in the first place, the economics of providing floodplain inundation would likely have been more favorable.

Ecological problems can also arise when flow releases change rapidly up or down (called "ramping"). Ecologically damaging ramping occurs when a dam suddenly begins spilling high volumes of water during a flood or when substan-tially greater volumes of water are released when additional turbines are activated. This can lead to high mortalities in fish and other animals in the river or on the floodplain or cause undesirable erosion and sedimentation problems downstream. Conversely, when releases from a hydropower dam are being reduced for the purposes of rebuilding water levels (head) in a storage reservoir by shutting down outlets, river flows can be curtailed too abruptly and leave less mobile animals such as mussels and small fish and their eggs high and dry at the river's edge. Providing a gradation in turbine-generator sizes and reservoir outlets in a dam's design will minimize problems with these flow transitions. Further, construction of "reregulating dams" downstream of a hydropower dam can catch and partially even out fluctuations by releasing water in run-of-the-river fashion.

When designing the outlet and turbine-generator capacity of a new dam, it is highly desirable to incorporate a wide range of water-release capabilities as well as adequate transmission capacity to convey the electricity, such that the full array of dam-operating objectives, ranging from hydropower generation to environ-mental flow releases, can be accommodated. By providing a range of outlet sizes, such as by incorporating multiple turbine-generator units of varying sizes, dam operators will be able to meet a variety of dam-operating objectives.

Multilevel (Selective Withdrawal) Outlet Structures

The water in many reservoirs can become stratified, with considerable differences in water temperature with depth in the reservoir. Water near the bottom of a reser-voir may contain very little dissolved oxygen, and this anoxic condition can cause chemical reactions that lead to undesirable water quality conditions in deep-water zones. The release of this water with low oxygen levels and undesirable chemicals can create serious problems for fish and other aquatic animals downstream of the dam. Multilevel outlet structures (also called "selective withdrawal" structures) can be constructed to provide dam operators with the flexibility to release water from different reservoir levels, depending on the time of year, differences in water quality and temperature, and downstream management objectives.

Reregulation Reservoirs

The impacts of hydropower generation on natural river flows can be mitigated to some degree by constructing a "reregulating" dam, usually built immediately downstream of the lowest hydropower dam. The reregulating dam can be operated to "smooth out" the unnatural fluctuations caused by hydropower operations even while it is generating electricity, releasing water in a pattern much closer to reservoir inflows. The ability of a reregulating dam to restore natural flow patterns will depend on the extent to which the upstream hydropower dam has altered them; essentially the same volume of storage capacity is needed both to alter flows at the hydropower dam and to restore flows at the reregulating dam. If hourly downstream fluctuations are undesirable, a relatively small reregulating dam below the powerhouse can be a positive asset to hydropower and to the environment by providing a more steady downstream discharge during the day. However, if a large reservoir is being used to reshape the hydrograph over several months, that same volume of storage would be needed in the reregulating reservoir to reshape the hydrograph back to a more natural pattern. The same benefit can be achieved by dedicating the lower-most hydropower dam in a cascade to reregulate flows, which can be of considerable benefit to the downstream environment.

Other Design Infrastructure

While provision of adequate environmental flow releases will go a long way toward maintaining adequate habitat conditions and ecosystem services in rivers affected by dam development, other ecosystem protection measures may be needed.

Sediment Bypasses and Sluice Gates

Sediment trapping in reservoirs presents serious challenges for water storage by reducing available storage capacity and creating the risk of uncontrolled dam overtopping and collapse. Sediment moving through a reservoir and into hydropower intakes can severely damage turbines and shorten their lifespan.

It may also disrupt geomorphic processes that create high-value habitat below the reservoir, especially in watersheds with high sediment production rates. If the water being released from the dam retains sufficient ability to erode the downstream river channel and banks, but sediment is not available from upstream to replace the eroded sediment, considerable channel down-cutting and instability can result, thereby endangering structures such as roads, bridges, and levees and altering the physical habitats supporting aquatic life. The loss of sediment supply to downstream deltas and coastal areas can result in considerable erosion of beaches and islands of great importance for people and nature as well.

The implementation of sediment management measures—in the contributing watershed and in the reservoir—can greatly extend the design life of a dam and lead to other economic benefits, such as reducing the costs of maintaining hydropower turbines. Passing sediment around or through a dam can also help to alleviate dam-related impacts in the downstream river ecosystem. The World Bank's publication on the RESCON software (Palmieri and others 2003) provides several useful approaches for managing sediment and evaluating the cost-effectiveness of sediment management measures. Some of the approaches discussed in that publication are summarized here.

In addition to providing dead storage to accommodate sediment deposition in a reservoir, new dam designs are including features to move sediment around or through the reservoir. These features are generally of two types. Sediment bypass structures are designed to route sediment inflows into a bypass outlet (a channel upstream of or in a reservoir that bypasses the dam and rejoins the river below the dam) and subsequently discharge sediment and water below the dam, thereby keeping sediment from flowing into hydropower turbines. Sediment flushing involves opening sediment sluice gates or other low-level outlets and lowering reservoir levels to cause water in the reservoir to begin to flow through the reservoir and outlets. This flow needs to attain sufficient velocity to flush the sediments that have accumulated in the reservoir. This type of reservoir flushing entails considerable tradeoff with power generation, however, because the reservoir level (head) must be lowered considerably, thereby compromising potential power generation during flushing. It can also complicate environmental flow management due to the fact that reservoir storage must be refilled following sediment flushing, reducing downstream flow releases during refill. Moreover, in large reservoirs, the sediment tends to deposit at the inflow end of the reservoir rather than behind the dam, limiting the ability to flush it through the sluice gates.

Fish Passage Structures

Structures such as "fish ladders" have commonly been used to enable fish and other mobile aquatic organisms to move upstream and downstream of a dam. However, the higher the dam wall, the harder and more expensive it is to build effective fish passages. Every dam, including those with fish passage structures, is likely to block the passage of some portion of the migratory fish. Each species will have particular design requirements for successful passage. For example, until recently, Australian dam builders constructed "horizontal baffle" fish ladders suitable for jumping salmonid fish (that is, trout, salmon) imported from the Northern Hemisphere. However, most of Australia's native fish do not jump and did not use these fish ladders, requiring instead "vertical slot" fish ladders that allow these species to rest in eddies at each step. Aquatic wildlife may migrate along river banks, requiring passages on each side of a barrier, or follow the "scent" of a strong water flow, requiring a strong current to flow from the wildlife passage to attract the animals to the entrance. Rock ramp fishways that mimic natural waterways may be the most effective wildlife passages, whereas at the other end of the spectrum, fish lifts and "catch and truck" operations are likely to assist only a modest portion of the migratory animals. Any dam without wildlife passage is likely to have a severe local impact on species diversity.

Background to Environmental Flows

THE FLOWS NEEDED TO MAINTAIN important ecosystem services are termed environmental flows. This report has adopted the following definition from the Nature Conservancy (2006):

> Environmental flows are the quality, quantity, and timing of water flows required to maintain the components, functions, processes, and resilience of aquatic ecosystems that provide goods and services to people.

This definition is general enough to include water for both surface water and groundwater systems and focuses on the need to maintain downstream ecosystem services valued by people. Water can be provided for environmental outcomes in two broad ways: (1) by specific releases of water from water storages or (2) by restrictions on abstractions from water systems. The former is termed "active management," and the latter is termed "restrictive management" (Acreman and Dunbar 2004). The former approach is usually only possible for regulated river systems where reservoirs are holding environmental water allocations;[1] the latter approach is more broadly applicable to both regulated and unregulated systems as well as to groundwater. It also recognizes the importance of ecosystems in providing services (use values) to people.

There has been a gradual increase in acceptance and integration of environmental and social issues into decision making about dams. Initially, decision making was under the domain of engineers alone, but over the last four decades it has expanded to include economists, environmental specialists, social scientists, and upstream displaced people. Most recently, it has included greater attention to downstream ecosystems and communities (see figure C.1).

Environmental flows are particularly important in river system management. Although diffuse and point-source pollution, introduction of exotic species, riparian degradation, and removal of aquatic habitat can all affect the provision of ecosystem services, the flow regime is of central importance because so many ecosystem functions depend on it. One aquatic ecologist has likened flow to "the maestro that orchestrates pattern and process in rivers" (Walker, Sheldon, and Puckridge 1995, quoted in Postel and Richter 2003). Flows influence aquatic ecosystems in four primary ways (Bunn and Arthington 2002):

- *Shaping physical habitats such as riffles, pools, islands, and bars in rivers and floodplains.* Alteration of flows can lead to severely modified channel and floodplain habitats, thereby affecting the physical diversity needed to support diverse aquatic communities.
- *Affecting life-cycle processes.* Many aquatic species depend on specific water flow conditions during life stages such as reproduction.
- *Altering mobility of organisms.* Many species need to move upstream and downstream or from the river to the floodplain during their life cycle. Flow alteration

FIGURE C.1
The Evolution of Dam Planning Practices

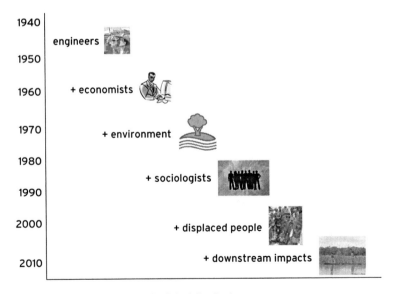

Source: Modified from a chart prepared by Robert Goodland.

that impairs connections between these different habitat areas can limit their mobility.
- *Creating conducive conditions for the invasion of exotic and introduced species.*

Other terms are sometimes used instead of "environmental flows." Some of these alternative terms reveal limitations in earlier conceptions of environmental flows:

- *Bypass flows* are used to describe the releases made from the dam into the lower Kihansi gorge ecosystem.
- *Escapages* are used to describe the Indus River flow at Kotri Barrage to check saltwater intrusion, accommodate fisheries and environmental sustainability, and maintain the river channel.
- *Minimum flows* are used to describe the retention of enough flow to maintain river connectivity, especially for fish passage, but this is usually only one component of the flow regime that needs to be maintained, and there are few instances where an environmental flow consists of just a minimum flow.
- *Instream flows* imply the flows needed to maintain ecosystem services from flows within the river channel, but this excludes the often important floodplain flows that overtop the channel.
- *Environmental water allocation* is used to describe water specifically allocated for environmental purposes, usually held in a dam or storage.
- *Ecological reserve* was coined in South Africa to describe the water allocated for downstream ecological functioning;[2] however, the term *reserve* implies water held back in reservoirs or impoundments, whereas environmental water can also be assigned through controls on abstractions and discharges.
- *Natural flows* imply that the environmental flows need to mimic the natural flow variability; in fact, environmental flows can deviate significantly from the natural flow regime when some flow components are maintained because of their important functions and other components are lost because they are considered to be unessential to the river ecosystem and can therefore be assigned to some development needs.
- *Surplus water* implies that some components of the flow regime have no ecosystem value and are available for assignment to consumptive or other purposes at no cost to the environment; in reality, all components of the flow regime provide some ecological function. Even flows to the ocean are not wasted water; they provide nutrients to estuaries and near-shore areas, trigger spawning in some fish and invertebrate species, and shape physical habitat.
- *Compensation flows* are used in the 1986 LHWP treaty between South Africa and Lesotho. The term is confusing since it implies that the remaining downstream flows are actually a compensation for the benefits that have been lost by abstracting the rest of the flows.

However, none of the terms, including *environmental flows*, gives sufficient prominence to the social and economic importance of these flows. There is thus a misleading implication that only the environment, and not the people dependent on the environment, benefits from environmental flows.

Scientific knowledge is central to making decisions on water allocation at either the basin plan or project level. Scientific knowledge provides reliable information on the predicted response of aquatic ecosystems to changes in the quantity, timing, and duration of flows; this is particularly important because the response of ecological systems to changes in flow is seldom linear. Thus reducing a component of the natural flow regime will not necessarily have a proportionately reduced ecological outcome. Halving the peak flood flow will not transport half the sediment, and halving the overbank flow will not usually inundate half the floodplain (Gordon and others 2004). Unless these critical thresholds are understood, there is a possibility that progressive changes to the flow regime will have little apparent effect until the threshold is reached, and then there will be an abrupt decline in ecosystem function.

Scientific knowledge is also important for understanding surface water hydrology and groundwater hydrogeology, the changes in water availability as a result of climate change, links between water quality and changes in flow, and the effects of changes in land use on runoff characteristics.

Scientific knowledge about ecosystem responses is not always available entirely or in adequate form or quantity, particularly in developing countries. Traditional knowledge about the responses of vegetation, fish, or birds to different quantities and timings of flow can provide valuable information (see box C.1) and can be used to supplement limited data and scientific knowledge.

Downstream individuals and communities who are affected by changes in flow regimes are often relatively unorganized and powerless compared to groups who want to develop the water resource. In addition, their traditional rights to use water are not always recognized in law. It is important for the relationships of the communities to rivers and the needs of these downstream communities to be included in decisions about flows through the following:

- Making their objectives and reliance on flows explicit
- Quantifying the links between different flow components and these objectives
- Involving them in the decision-making process when choices are made about modifying components of the flow regime.

Most EFA methods have focused on assessing biophysical impacts. It is only recently that EFA methods, such as the DRIFT method developed under the Lesotho Highlands Water Project (case study 14), are attempting systematically to quantify downstream social impacts associated with biophysical impacts (King, Brown, and Sabet 2003). Methods such as DRIFT identify the reliance

---

BOX C.1
Using Indigenous Knowledge, Rio Patuca, Honduras

With most of the nation's hydropower potential still undeveloped, Honduras has decided to construct a hydropower dam on the Rio Patuca, the country's longest river and the third longest river in Central America.

The Patuca's lower reach passes through a region of immense cultural and biological value. The river provides important ecosystem services to Tawaka, Miskito, and Pech communities along the banks of the river. Fisheries serve as their major source of protein; sediment deposition during annual flooding improves the fertility of low-lying agricultural fields; and the river is their primary means of transportation. The Empresa Nacional de Energía Eléctrica, the Honduran energy agency, asked the Nature Conservancy to provide guidance on a flow regime below the proposed dam that would maintain the river's biodiversity and ecosystem services.

Due to a paucity of technical information, the Nature Conservancy used traditional ecological knowledge about fisheries, agriculture, and transportation derived from the communities along the river as the basis for flow recommendations. Community members provided information on flow levels through two sources of spatial information: cross-sectional surveys of the river and historical watermarks and hand-drawn maps of each community. This was synthesized with other regional information to develop conceptual models of the linkages between flows and important fish species. These sources of information, augmented by hydrological analysis of a 30-year record of daily flows, provided the foundation for the environmental flow assessment.

Source: Nature Conservancy and Natural Heritage Institute forthcoming. http://www.nature. org/initiatives/freshwater/files/final_patuca_case_study_low_res_new_logo.pdf.

---

of communities on flows and quantify the links between flows and objectives. However, the extent to which stakeholders are involved in decisions about water allocation is determined by water (and sometimes environmental) policy and legislation as well as by power structures and customary law. In many developing countries, downstream communities (including those with customary water rights) have no tradition of being involved in decisions and constitute voiceless constituencies (Hirji and Watson 2007).

Notes

1 There are special circumstances where water can be purposefully released into a water system as return flows from urban and irrigation uses, but these are difficult to manage for environmental outcomes and are seldom used.

2 In South African law, the ecological reserve is distinguished from the social reserve, which is water allocated for basic human needs (case study 3).

Water Environmental Issues in Country Water Resources Assistance Strategies

THIS APPENDIX PRESENTS water issues in the World Bank CWRASs of 17 countries or regions: Bangladesh, China, Dominican Republic, East Asia and the Pacific region, Ethiopia, Honduras, India, Islamic Republic of Iran, Iraq, Kenya, Mekong region, Mozambique, Pakistan, Peru, the Philippines, Tanzania, and Republic of Yemen.

Bangladesh

Bangladesh, more than most countries, depends on water-derived ecosystem services for its survival. The Bangladesh CWRAS reflects this with recognition that the country is heavily dependent on the Ganges and Brahmaputra rivers. Fisheries production is under threat from reductions in dry-season flows, loss of key aquatic habitats, and disruption to migratory pathways (80 percent of rural Bangladeshis depend on aquatic resources); there are water shortages in the southwest; the Sundarban ecosystems are deteriorating; and water transport has contracted because of reduced flows (from 8,500 kilometers of navigable waterways in winter in 1970 to 3,800 kilometers today).

These downstream effects arise partly from developments in both Bangladesh and India. The Ganges water treaty provides a stable framework within which Bangladesh can begin to plan for the development of the main rivers. Specifically, it provides some assurance of upstream discharges so that the Gorai River Augmentation Project could be implemented with some promise. There are risks.

The Gorai augmentation study did not adequately deal with the possibility that the Ganges discharge at the Gorai offtake may be too low. This reduces the volume of water available for redirection into the Gorai. A second area of concern is that the operation of the gates at Farakka has led to a steeper recession limb of the hydrograph. As a result, there are more residual bars and pools that restrict free channel flow.

The CWRAS proposes a strategy for Bank engagement that includes a full-scale assessment of human and environmental impacts of development on the Ganges River, establishing a scientific basis for determining environmental flows. The CWRAS says that the scientifically based assessment of environmental flows is necessary to promote understanding between the upper and lower riparian countries to help develop benefit-sharing agreements for the river.

China

The CWRAS bluntly says that China is not maximizing its benefits from Bank involvement. It should make greater use of the Bank's technical capacity. One of the recommendations is that China should use the Bank's financing for management of water environmental issues such as ecosystem restoration of rivers, wetlands, lakes, and coastal waters.

The CWRAS notes that overexploitation of groundwater, particularly in the Hai basin, and overuse of surface water resulting in inadequate environmental flows in much of northern China, along with increasing groundwater and surface water pollution in many parts of the country, are contributing to the decline and deterioration of water resources and damage to freshwater and coastal environments. More broadly, China needs to protect and restore the environment; otherwise, environmental degradation will result in huge negative impacts on the quality of life of the Chinese people.

The CWRAS proposes six themes for assistance, two of which are relevant to environmental flows. The first theme is to improve environmental water management. This includes national guidelines for comprehensive river basin planning, which include a requirement for environmental flows in rivers providing water to ecologically important areas. The Bank can assist in this area. The sixth theme advocates further investment in new water resources infrastructure. Although there is a high priority accorded to rehabilitating environmentally degraded water bodies, the CWRAS does not mention that this new infrastructure needs to be designed to protect downstream environments.

The 2002 Water Law contains provisions for allocating water for ecological and environmental protection and restoration. The CWRAS proposes that the Bank assist in revising the laws and in implementing them. The CWRAS strongly advocates much greater attention to the provision of water for environmental flows and proposes that China draw on the Bank's expertise in this area.

Dominican Republic

The Dominican Republic has severe problems related to water quality, flooding, and watershed degradation. There is no water resources policy or strategy, and the water law needs revision. The CWRAS focuses on addressing these issues and proposes that the Bank can assist with watershed protection as well as with revisions to the general water law. Although the CWRAS notes that there are water allocation issues, there is no mention of downstream impacts from upstream development apart from sedimentation from watershed erosion.

The country is seeking funding for construction of infrastructure, including dams, but there is no mention of the steps to be taken for controlling their upstream and downstream impacts.

East Asia and the Pacific Region

This regional CWRAS clearly recognizes the downstream impacts of upstream water uses. It provides examples, including the Tarim basin (case study 17) and the Mekong basin (case study 7), where environmental flows have been included in development assistance.

The CWRAS accepts that the Bank will reinvest in water resources infrastructure in the East Asia and the Pacific region, but questions the legitimacy of government initiatives and processes, including adequate consideration for the environmental impacts. It describes the shift of benefits from downstream water users to those upstream or offstream without using the language of environmental flows.

The CWRAS proposes seven themes for assistance. The first is to preserve the environment and the base of land and water resources. This requires the development of basin water resource plans that include environmental flows to protect rivers and coastal zones and to sustain ecologically important areas. This has not historically been done in the region, but this is now changing, as shown by the Tarim basin and, to a lesser extent, the Mekong basin examples. The third theme is for rehabilitation of existing water resources infrastructure and construction of new infrastructure. Although the need to provide for environmental flows is not specifically mentioned in this theme, it is clear that this development has to be accompanied by improved management and protection of environmental and social needs.

Ethiopia

The CWRAS includes very little mention of environmental flow issues. It does state that urbanization, industry, and services all create additional demand for water, potentially diminishing both the quantity and quality of water. Abstraction for these purposes will lower flow levels and have severe negative impacts on downstream users and the environment.

The CWRAS, in line with the energy strategy and the water sector development plan, advocates an expansion of hydropower and multipurpose development, but makes only passing mention of deleterious downstream effects: "Water releases for power generation must be weighed against requirements for irrigation, all in line with system requirements such as environmental flows."

Honduras

The Honduran CWRAS is yet to be completed. The draft CWRAS notes that water resources infrastructure is needed to meet MDGs and that this infrastructure development needs sound environmental management practices to ensure its sustainability. However, the country lacks the capacity to ensure that these developments are environmentally sustainable and does not have a water resources policy or up-to-date water law. The CWRAS does not comment on the need to include environmental flow considerations into project assessments or into new water resources policy and law.

India

Although environmental protection and restoration are important in Indian water management, the CWRAS focuses primarily on governance and institutional issues. The Indian state water apparatus is still focused on a command-and-control approach and has yet to shift to a modern approach based on incentives, participation, devolution, and environmental sustainability. Consequently, the CWRAS says, water managers ignore the accumulated "environmental debt" (including vanishing wetlands and polluted rivers and aquifers).

The CWRAS accepts that more water storage is needed to provide security, especially in the face of climate change, but this should be accompanied by more responsible management. However, dams are still seen within India as a solution, without the understanding that they can solve one person's problem at the expense of someone "downstream." The CWRAS argues that new investments need to be accompanied by greater care to safeguard existing downstream uses. Attention also needs to be paid to improving the reliability of supplying existing demands and meeting historically deprived environmental uses.

Unlike other CWRASs, the document does not present a strategy for Bank engagement with the water sector. Instead it provides 12 rules to guide the government of India in its water management. The India CAS already contains a major increase in water resources assistance. In discussions leading up to the CWRAS, there was a strong endorsement of the Bank's reengagement in the full range of water-related issues and agreement that the government needed to complement its traditional focus on infrastructure with a growing emphasis on management.

Islamic Republic of Iran

The CWRAS describes the Iranian water management as "unevolved." Although both water policy and law require that water be managed at the basin level, the number of basin organizations has not yet been established. Consequently, there is no call yet for basin-level water allocation planning or environmental water allocations.

However, the CWRAS says that there is a recognition that construction of dams will lead to less water for aquatic ecosystems. There are no guidelines or requirements at present for protecting downstream environments, but the CWRAS proposes that social equity and environmental criteria should be established to mitigate the negative impacts of water resources development.

The CWRAS identifies several areas where the Islamic Republic of Iran needs training and capacity building, including in understanding water and environment.

Iraq

The Iraq CWRAS contains a significant section on the loss of environmental flows, primarily in the Euphrates River, which has contributed to the devastation of the Mesopotamian marshes at the confluence of the Euphrates and Tigris rivers.[1] About 94 percent of the Euphrates flow comes from Turkey, with the remaining 6 percent coming from Syria. Both countries have developed large hydropower dams in the headwaters that have intercepted more than 50 percent of the flows. The planned irrigation developments will intercept even more. The CWRAS points out that the full irrigation development will have major impacts on water quality as well as leading to further environmental degradation in the marshes. Environmental flows are not just a matter of quantity; timing is also important for marsh inundation, as large volumes of water are required in a concerted surge to flood the marshland.

The CWRAS recognizes the need for instream flows for both strictly environmental and navigational needs, although these environmental flow needs are not elaborated. Iraq's hydropower generation lies on the Euphrates River and so is directly linked to the transboundary agenda. These issues require a transboundary technical process to manage flows optimally at the basin scale.

Much of the country's water resource infrastructure is degraded and even dangerous. The rehabilitation of these assets is identified in the CWRAS as a priority. There are no plans for constructing new dams. However, there is no mention of the opportunity for establishing environmental flows below rehabilitated dams, even though the importance of these flows is clearly apparent in the transboundary case.

Finally, the CWRAS includes the development of new water policy and governance arrangements as a priority, but there is little detail on the content of the new policy.

Kenya

There is little mention of environmental flow issues in the Kenyan CWRAS apart from recognition that existing hydropower dams and at least one irrigation scheme are causing downstream environmental and social problems. The issues include reduced fish catches, loss of spiritual value, and danger from unannounced high flows. This is the result of poor infrastructure planning. The CWRAS calls for upstream and downstream communities to be more engaged in preparation for new developments.

Kenya's 1999 water policy identifies catchments as the basic unit for water allocation planning and management. However, this intention has not been translated into practice. The 2002 Water Act also identified the need for a reserve to safeguard basic human and ecosystem needs. However, it too has not been implemented.

Mekong Region

This transboundary CWRAS says that at present there is inadequate coordination of development in the Mekong region, with consequent social and environmental risks. Thus there is great momentum for hydropower development; this will reduce floods and increase dry-season flows. Mainstem dams on the lower Mekong will impede fish movement. However, the document notes that it is possible to have sustainable development of the Mekong's water resources, while avoiding or minimizing negative impacts on the interests of other riparian countries and on important environmental and social values.

Although the environmental flow rule was reduced in status to a guideline (case study 7), the CWRAS sees the intense interest during this debate as a positive sign of engagement over the topic of environmental flows.

The Mekong Water Resources Partnership Program Action and Dialogue Priority Framework (which the CWRAS supports) should include projects at subregional scale that promote, among other things, environmental programs and that mainstream environmental and social safeguards. Overall, the CWRAS identifies potential downstream environmental effects of development, but it does not recommend specific actions to avoid these problems.

Mozambique

The CWRAS recognizes the importance of providing an ecological reserve and calls for environmental water requirements to be established in each river basin. However, the major environmental flow issue (and opportunity) facing Mozambique arises from the downstream effects of the Cahora Bassa Dam, which was established in 1975 without any provisions for environmental flows. There has been a decline in downstream fisheries, major changes have occurred in the Marromeu wetlands and other delta and riverside forests, flood-recession agriculture can no longer be practiced, and the estuarine prawn fishery has declined. The CWRAS notes that the change in ownership of the Cahora Bassa Dam opens up the opportunity to modify the operations of the dam so that downstream environmental and social considerations are taken into account.

New dams are proposed on the Pungue River to provide water supply to Beira and on the Zambezi River at Mphanda Nkuwa and Cahora Bassa North for multiple purposes. The Pungue Dam needs to provide downstream flows to prevent saltwater intrusion into the delta, with the new developments on the Zambezi River operated to provide the flows needed to reestablish downstream ecosystem services.

The CWRAS includes the need for Mozambique to build its capacity for integrated water resources management, including hydrological and environmental monitoring.

Pakistan

Although salinity is seen to be the major water-related environmental threat in Pakistan, the CWRAS also identifies flow-related issues as being important, particularly in the delta of the Indus River. The delta has become degraded from several causes, including the reduction in freshwater outflow and the decrease in accompanying sediments and nutrients. The CWRAS states that "it is important to provide some managed flows to sustain the delta to the degree that this is possible," although studies were yet to be completed at that time to determine how much flow was required.

Wetlands are also under severe threat with reductions in flood flows; upstream infrastructure at Tarbela and Mangla are a primary cause. Embankments have now cut floodplains off from their rivers, with the consequent loss of buffering capacity during times of flood.

There is also need to rehabilitate a large stock of old infrastructure and to build new infrastructure. However, there is no mention of protecting or restoring downstream environments during this infrastructure investment.

The CWRAS pragmatically focuses on the major issues facing Pakistan. Improving flows to restore the ecosystems of the Indus estuary is the only activity where the document explicitly advocates environmental flows.

Peru

The CWRAS does not mention environmental water issues. Although it states that a set of environmental policies were approved in 2002, it is not clear what they included. The major water issues facing Peru include health-rated water quality issues, access to sufficient water supply in dry coastal areas, and institutional reform. Environmental water issues do not appear to be seen as important.

The Philippines

The CWRAS mentions at an early stage that the Philippines faces "overexploitation of groundwater (particularly in and around the larger cities) and overuse of surface water, resulting in inadequate environmental flows for major basins and sub-basins." There is a strong focus in the document on the need for water resources planning, development, and management to provide for ecological protection and sustain the environment. This planning should make provisions for environmental flows in rivers and coastal zones and for sustenance of ecologically important areas. The environmental water needs of aquifers should also be determined.

Although the document also states that there is a need for both new water resources infrastructure as well as better management of existing infrastructure, the specific requirements are not spelled out and there is no mention of the need to ensure that environmental flows are incorporated into the planning of these projects.

The Bank was assisting with the River Basin and Watershed Management Project at the time of the CWRAS. The CWRAS proposes that Bank-supported river basin management projects should include environmental sustainability components, including environmental flows for riverine and coastal benefit.

Tanzania

The Tanzanian CWRAS describes a number of occurrences where the neglect of environmental flows during agricultural and hydropower developments has led to downstream issues, especially in the Rufiji basin. The 2002 national water policy includes a requirement for environmental flows. However, the CWRAS states that there are no standards or guidelines for establishing environmental flows, although a program has been designed (but has yet to be implemented) to train water resources staff in environmental flow assessments and implementation.

The CWRAS identifies the need to include training in environmental aspects of water planning and management as an area where the Bank could assist. There is also a need to build an understanding among sectoral agencies of the importance of environmental flows.

Republic of Yemen

The Republic of Yemen has been reforming its water sector since the mid-1990s in response to water shortages. The lack of control over groundwater use had led to shortages for both agricultural and urban uses. There is considerable concern over the lack of physical sustainability of groundwater at the current rates of use, but there is no mention of any environmental impacts from the drop in water tables. Surface catchment problems are focused on the evident deterioration of watersheds, with impacts on downstream communities from upstream abstractions and pollution. Environmental issues do not figure in this description.

The CWRAS critiques the Yemen National Water Sector Strategy Investment Program and identifies a number of omissions and weaknesses. None of these includes environmental sustainability issues. The CWRAS advocates that the Bank should invest in a watershed management project to balance upstream and downstream water uses and should support basin planning within the current framework.

The absence of any mention of environmental flow issues for surface water or groundwater probably reflects the much higher priority being accorded to physical sustainability and equity issues.

Note

1 The Mesopotamian marshes now occupy only about 10 percent of their original area. Apart from the upstream impoundments and diversions, the Sadam Hussein government undertook deliberate engineering works to drain the marshes in retaliation for a revolt by the marsh Arabs following the First Gulf War.

Environmental Flow Programs of International Development Organizations and NGOs

SEVERAL PROMINENT INTERNATIONAL organizations, conservation NGOs, and research organizations[1] have offered assistance to developing countries seeking to address and undertake EFAs and protect and restore downstream ecosystems. The assistance includes practical, longer-term technical assistance with EFAs for specific infrastructure projects, technical assistance and financial assistance for efforts to include concerns about downstream flow into river basin plans, shorter-term training and capacity building, and provision of resources for water resource and environmental specialists.

The Bank is collaborating with development partners at several levels—global, regional, national, and basin—because of their experience, expertise, and comparative advantages as well as their presence on the ground. The Nature Conservancy and the Natural Heritage Institute have produced a technical guidance note for the Bank on integrating environmental flows into hydropower dam planning, design, and operations as a contribution to this ESW (Nature Conservancy and Natural Heritage Institute forthcoming). This has been published as a standalone note. The Bank is also collaborating with NHI and GEF to explore opportunities for examining the feasibility of reoperating existing dams and water systems in order to improve their economic and environmental performance. The relevant environmental-flow-related work of the various agencies is described in this appendix to inform Bank staff of the types of activities and potential opportunities for future collaboration. It is not intended to evaluate the effectiveness of the various EFA programs of these agencies.

The majority of the assistance offered by international agencies and NGOs is for EFAs for river basin planning and project assessment, particularly reoperation of existing infrastructure, and the provision of training courses. Some of these institutions have also developed a wide range of support material, including printed and electronic documents, databases, a newsletter, training courses, Web sites, and an information network.

Although the World Bank has worked with some of the specialist international NGOs in developing technical advisory documents, there are opportunities to increase the level of collaboration to combine their experience in EFAs and in training with the Bank's experience in implementing projects. Contact details of these various partner agencies and NGOs are provided at the end of this appendix.

International agencies and NGOs have been active in providing assistance to developing countries for incorporating environmental flows in policy reforms, basin and catchment plans, and assessments of new and rehabilitated infrastructure projects.

Policy and Legislation

A small number of international NGOs have influenced policy and legislation. In the eastern Himalayas in India, WWF is carrying out an environmental flow scoping study that will provide a platform to promote integration of the concept of environmental flows at the policy level. IUCN introduced the concept of environmental flows to the environment committee of the Costa Rican Parliament at the time a new water law was being proposed. Members of an expert network established by IUCN have subsequently been involved in preparing the new law. NHI was an author of a statutory mechanism in California for reallocating existing irrigation and municipal water rights to environment flows. The Environmental Water Account—a "water district for the environment"—is now the largest water purchaser in California.

Basin Planning

Various international agencies and NGOs provide assistance in incorporating environmental flow considerations into river basin plans.

IUCN, through its Water and Nature Initiative (WANI), is funding case studies in Mesoamerica, Southeast Asia, and Eastern Africa, including an EFA in the Pangani basin, Tanzania (case study 8). The EFA will contribute to preparation of the basin water resources plan required under the draft Tanzanian water resources legislation and is helping to build capacity in EFA within Tanzania. IUCN has also conducted a demonstration EFA in the Tempisque River basin in Costa Rica, which has stimulated interest in undertaking an EFA in Costa Rica's Savgre River basin (Jiménez and others 2005).

WWF is working with the USAID-funded Global Water for Sustainability (GLOWS) Program in the Mara River basin, transboundary to Kenya and Tanzania, to help ministries in both countries to carry out an EFA in preparation for basin planning. The EFA includes the environmental flow needs of the Masai Mara National Reserve and the Serengeti National Park, particularly during the dry season. In addition, WWF and DANIDA have initiated an EFA in the Ruaha basin in Tanzania. WWF has also identified a suitable methodology and developed an action plan for estimating environmental flow requirements in the Neretva River basin in Bosnia-Herzegovina.

UNDP, as a GEF-implementing agency, is planning to conduct EFAs in two international river basins. One is in the Orange-Senqu River basin (Botswana, Lesotho, Namibia, South Africa), and the other is in the Okavango River basin (Angola, Namibia, South Africa). See box E.1. The EFA will be conducted as part of the transboundary diagnostic analysis required in GEF international waters projects. UNDP has also been working on the Zarka River basin in Jordan, looking at issues of environmental flows as part of the introduction of integrated river basin management concepts.

The UNEP Global Programme of Action has developed a protocol for carrying out EFAs in Bangladesh. The protocol has been piloted at the Bakkhali River Rubber Dam, where there was a need to establish a balance between water for irrigation and dry-season flows for fish movement to increase fish production.

IWMI has also been active in promoting EFAs within river basins, particularly in Asia. It is working with IUCN and Vietnamese government agencies to develop an environmental flow program in the Huong River basin in Vietnam (IUCN Vietnam 2005). To date, the assistance has been restricted to raising awareness and introducing concepts. IWMI has also applied the Nature Conservancy's range of variability approach (Richter and others 1996) to three rivers in the East Rapti River basin, Nepal (Smakhtin, Shilpakar, and Hughes 2006), and to the Walawe River basin in Sri Lanka (Smakhtin and Weragala 2005). However, the lack of ecological data (in the Nepal applications) and the high uncertainty in the estimation of the natural flow regimes have limited the applicability of the method. IWMI has also applied the South African desktop method (case study 3) to assess the environmental flows of rivers in India, including the Cauvery, Krishna, Godavari, Narmada, and Mahanadi rivers (Smakhtin and Anputhas 2006).

These applications illustrate a general lesson: it is difficult to transfer a method from one region of the world to another without carefully considering the assumptions behind the method and understanding the purpose for which the method was developed and its limitations.

BOX E.1
Flows in the Okavango Basin

The Okavango basin, covering parts of Angola, Botswana, and Namibia, is one of the most natural areas on the African continent. The Okavango River forms a large inland delta, which comprises a large perennial swamp, a seasonally flooded swamp, seasonally flooded grassland, intermittently flooded land, and drylands. The complex includes more than 150,000 islands varying in size from several meters to 10 kilometers in length. This flood-pulse system is driven largely by rainfall in the upper parts of the catchment in Angola.

The delta provides unique habitat that supports a rich and diverse biota, including some of Africa's largest free-roaming herds of Cape buffalo, zebras, antelope, and elephants. The delta also includes between 2,000 and 3,000 species of plants, more than 65 species of fish, more than 162 arachnid species, and more than 650 species of birds. The associated tourism-related activities are the second most important economic activity in Botswana, after diamonds.

Several potential sites for the development of hydropower generation have been identified in the upper reaches, and potential irrigation development has been identified in several parts of the basin. During the 1980s, the government of Botswana proposed a water development project to use water from the delta for mining, agriculture, and cattle production. The project was rejected as a result of a wide range of deficiencies relating both to the project itself and to the planning and design process.

However, the changing regional political context and the need to ensure socioeconomic development in some of the region's most remote and under-developed areas mean that there is continuing pressure to develop the waters of the delta. Consequently, there is a newfound urgency to secure environmental allocations to protect the delta's rich diversity.

Increasing development pressures, the need to address development challenges, and increasing peace and stability throughout the basin resulted in establishment of an institutional framework to facilitate joint planning. The three riparian countries established the permanent Okavango River Basin Water Commission (OKACOM) in 1994. This is a tripartite institution aimed at promoting coordinated and environmentally sustainable regional water resources development, while addressing the legitimate social and economic needs of each of the riparian states. The OKACOM Secretariat has responsibility for determining and facilitating water allocations within the Okavango basin. However, there has yet to be a rigorous assessment of the environmental flow requirements of the Okavango River basin.

Source: http://www.okacom.org.

IUCN has run a wetlands program since the early 1990s in conjunction with Ramsar. This work included the determination of the water needed to maintain ecological functioning of wetland habitats and the delivery of goods and services to local livelihoods. Although this work was not undertaken for basin planning purposes, it would contribute to basin-level water allocation planning. The term environmental flows was not used to describe this work, but it would be classified under this heading today.

Rehabilitating and Reoptimizing Infrastructure

NHI specializes in reoptimizing major irrigation, power, and flood management systems to add to the supply side of the water balance. (Appendix B contains a summary of the infrastructure design features needed for releasing environmental flows.) The reoperation techniques, which include economic optimization modeling, conjunctive management of surface water and groundwater, reduction of physical losses of water in irrigation, and rescheduling of total hydropower production and system reliability, provide more water to formerly productive downstream river systems in ways that do not significantly reduce production benefits. Demonstration activities are being considered in the Yangtze, Yellow, and Pearl River basins in China, in the Hadejia-Nguru wetlands system in northern Nigeria, and for the Akosombo and Kpong dams on the lower Volta in Ghana.

NHI is also helping to optimize surface water and groundwater management in the central valley of California to provide more water for the environment. It is also exploring a variety of economic incentives that can be used to improve agricultural water use efficiency, with the "saved" water being dedicated to the restoration of environmental flow.

NHI is working with GEF-implementing agencies to help developing countries to reoperate existing dams. It prepared a project concept to GEF for a project that will pilot the reoptimization of two dams, in Ghana and Nigeria, to enhance environmental flows as well as power generation and water-related livelihoods.

The hydrology of the Parana River, Brazil, has been severely altered by 26 large reservoirs. UNESCO, through its Ecohydrology Programme, is helping to restore the ecosystem functioning of this river through a modified procedure for operating the Porto Primavera Dam. This will maintain the biodiversity of a reach of the river and improve local incomes without significant loss of hydroelectric production.

The Nature Conservancy has worked extensively with dam operators in the United States to modify how and when water is released in order to restore and protect river systems and associated land and wetlands. In particular, its collaboration with the U.S. Army Corps of Engineers at 26 dams in 13 states across the United States has helped to define and implement environmental flows through adaptive reservoir management, and this experience is now being rolled out across other dams operated by the USACE. These studies have resulted in changes in the timing, rather than the volume, of flow releases, thus minimizing the cost to the operators of the dams.

In Mozambique, the Nature Conservancy is preparing to assist the Zambezi River Authority to rebuild the river's health by restoring environmental flows below the Kariba and Cahorra Bassa Dams.

WWF has worked with Zambian authorities to improve the operating rules of the Kafue Gorge and Itezhi-tezhi Dams in order to improve the management of water resources in the Kafue flats, which are a wetland of international importance under Ramsar. The aim is to provide a more natural flow regime in order to restore wetland functions and values.

New Infrastructure

NHI and the Nature Conservancy are helping to introduce environmental flows concepts in Africa, China, and Latin America. The Nature Conservancy has been developing a comprehensive conservation plan for the upper Yangtze, while collaborating with World Wide Fund for Nature, which is working on the lower Yangtze. As part of this, it facilitated a meeting, which has led to a draft report that lays the basis for environmental flows in the upper Yangtze. It has more recently been invited to assist the Three Gorges Company in developing environmental flow recommendations.

In Latin America, the Nature Conservancy has been conducting an environmental flow assessment for the Patuca III hydropower plant on the Patuca River in Honduras. The river supports globally important aquatic biodiversity and flows through a reserve for indigenous communities and other protected areas. The EFA will provide the information for protecting these important biological and cultural values. The Nature Conservancy's "payment for environmental services project" at Quito, Ecuador, includes an environmental flows component. The Nature Conservancy is also assisting with environmental flow analyses for rivers in Colombia and Peru, where new water fund projects are following the Quito model.

Training and Capacity Building

A considerable number of international agencies and NGOs offer support for training and capacity building in environmental flows. UNDP promotes capacity building at local, regional, national, and global levels through the Cap-Net Program implemented by UNDP together with the Global Water Partnership and UNESCO's Institute for Water Education. The program supports 12 regional and national networks of water management capacity-building institutions around the world. In South Africa, a Cap-Net network called WaterNet[2] is using its members to develop a regional master's program in water management. This includes training modules that focus on water for the environment to maintain ecosystem functioning.

In addition to conducting regular training courses for the USACE, the Nature Conservancy provides environmental flows training at international conferences so that participants from developing countries have opportunities to receive training in environmental flows.

Some NGOs have conducted awareness-raising workshops. IWMI has run a national workshop in India on environmental flows that brought together government departments, nongovernmental organizations, and research institutions. IUCN has organized workshops to raise awareness and increase understanding of environmental flows in Vietnam (see box E.2) and Cambodia. IUCN has also

---

BOX E.2
The Huong River Basin, Vietnam

The Tam Giang-Cau Hai lagoon at the mouth of the Huong River in central Vietnam is an important asset for local villagers. It provides fisheries and brackish water for agriculture, transport, and harbor facilities and, in recent times, has been used for aquaculture. It is also suitable for tourism development.

A number of dikes and barriers have been erected to prevent flooding along the river and dams; the most notable is the Thao Long Barrage. Barriers above the lagoon now prevent seawater intrusion from the lagoon during the dry season. Two new dams—Ta Trach and Binh Dien—have been proposed to provide further flood protection. Provincial officials were keen to develop an ecosystem-based approach to managing the basin and its infrastructure and asked IUCN and IWMI to assist. A rapid assessment EFA workshop was held in Hanoi in December 2004.

Prior to the workshop, Vietnamese water managers understood environmental flows to mean minimum flows, which were established through a hydrologic formula. They had set this minimum flow to be 31 cubic meters per second, to be released from the Thao Long Barrage to flow into the lagoon. In their view, it had become the dominant flow requirement.

The workshop went beyond the usual rapid hydrological assessment to include ecological assessments of different hydrological scenarios, including the effects of the proposed dams. While this rapid assessment was inadequate for establishing defensible environmental flows, it was successful in raising awareness about the use of a holistic approach to flow assessment. It also succeeded in identifying possible barriers to EFA implementation and ways to overcome these barriers. It will be important to link further work on environmental flows in this river basin to poverty alleviation and livelihoods.

Source: IUCN Vietnam 2005.

run a training course in Mesoamerica to develop a nucleus of champions of environmental flows.

Some training is available electronically. The Nature Conservancy will be making its training courses available online, and IUCN has established a distance learning electronic training course[3] that has modules designed for decision makers, which explain the benefits of environmental flows, and technical managers, which elaborate on methods for assessment, adaptations to existing or proposed infrastructure, and financing of EFAs.

Resource Materials and Awareness

There is a considerable array of electronic and hardcopy resource materials from international organizations and NGOs to assist developing countries (see box E.3).

The following resources are available electronically:

- A global environmental flows network, established by IUCN, IWMI, the Nature Conservancy, and several other development assistance institutions,[4] which

---

BOX E.3

Environmental Sustainability in Southern Africa

The Southern Africa Development Community—with support from the Southern Africa Research and Documentation Centre, IUCN, Swedish International Development Cooperation Agency, and the World Bank—has published a report on environmental sustainability in water resources management. It contains chapters on the role of aquatic ecosystems in water resources management, valuation of the environment, and the application of EFA in southern Africa. It is a valuable resource for environmental flows in that region.

The report identifies 10 challenges facing the introduction of EFA in southern Africa:

1. Lack of political will
2. Poor harmonization of policies for transboundary resources
3. Limited awareness and training in EFA
4. Lack of data on Southern African rivers
5. Inherent unpredictability of complex systems
6. Unknown influence of climate change on runoff
7. Lack of monitoring programs
8. Incompatible dam design with the necessary environmental flow releases
9. Need to rectify poor design of existing dams
10. Need to treat water as a finite resource.

Source: Hirji and others 2002.

acts as a central reference point for knowledge and information on environmental flows

- An environmental flows newsletter (IWMI and the Global Water Partnership)
- Three publicly available data sets developed by IWMI on (1) estimates of environmental flow requirements in world river basins; (2) environmental flow assessment for aquatic ecosystems; and (3) quantification of hydrological functions of inland wetlands
- A flow restoration database, which catalogues and organizes case studies of modified dam operations, removal of dams, groundwater pumping, and other strategies to restore river flows (the Nature Conservancy).

Finally, two international environmental flow conferences have been held, the first in Cape Town, South Africa, and the second in 2007 in Brisbane, Australia. A third is planned for South Africa in March 2009.

Contact Details

IUCN
Rue Mauverney 28
Gland
1196
Switzerland
PH: +41 (22) 999-0000
FAX: +41 (22) 999-0002
EMAIL: webmaster@iucn.org

IWMI
Headquarters
127, Sunil Mawatha,
Pelwatte,
Battaramulla,
Sri Lanka
PH: +94 (11) 288-0000
FAX: +94 (11) 278-6854
EMAIL: iwmi@cgiar.org

Natural Heritage Institute
Main Office
100 Pine St., Suite 1550
San Francisco, CA 94111 USA
PH: +1 (415) 693-3000
FAX: +1 (415) 693-3178
EMAIL: nhi@n-h-i.org

The Nature Conservancy
Worldwide Office
4245 North Fairfax Drive,
Suite 100
Arlington, VA 22203-1606 USA
PH: +1 (703) 841-5300
WEB: http://www.nature.org

UNDP
Headquarters
United Nations Development Programme
One United Nations Plaza
New York, NY 10017 USA
PH: +1 (212) 906-5000
FAX: +1 (212) 906-5364

UNEP
United Nations Environment Programme
United Nations Avenue, Gigiri
P.O. Box 30552, 00100
Nairobi, Kenya
PH: +254 (20) 762-1234
FAX: +254 (20) 762-4489/90

UNESCO
1, rue Miollis
75732 Paris Cedex 15
France
PH: +33 (0)1 45 681-000
FAX: +33 (0)1 45 671-690
WEB: www.unesco.org
EMAIL: bpi@unesco.org

WWF
WWF International,
Av. du Mont-Blanc 1196
Gland
Switzerland
FAX: +41 (22) 364-0074
WEB: http://www.panda.org/

Notes

1 The Center for Ecology and Hydrology (Wallingford, UK), University of Cape Town (South Africa), and Florida International University are three prominent research organizations that are active in providing assistance in environmental flows to developing countries.

2 http://www.waternetonline.ihe.nl/default.php.

3 Available at www.waterandnature.org/flow.

4 Stockholm International Water Institute, DHI Water and Environment, Centre for Ecology and Hydrology, Swedish Water House, and Delft Hydraulics.

References

Acreman, Michael C., and Michael J. Dunbar. 2004. "Methods for Defining Environmental River Flow Requirements: A Review." *Hydrology and Earth System Sciences* 8 (5): 861–76.

Arthington, Angela H., and Jacinta M. Zalucki, eds. 1998. "Comparative Evaluation of Environmental Flow Assessment Techniques: Review of Methods." LWRRDC Occasional Paper 27/98, Land and Water Resources Research and Development Corporation, Canberra, Australia.

Barbier, Edward B., William M. Adams, and Kevin Kimmage. 1991. "Economic Valuation of Wetland Benefits: The Hadejia-Jama'are Floodplain, Nigeria." London Environmental Economics Centre Paper DP 91-02, International Institute for Environment and Development, London.

Belt, George C. B. Jr. 1975. "The 1973 Flood and Man's Constriction of the Mississippi River." *Science* 189 (4204): 681–84.

Bunn, Stuart E., and Angela H. Arthington. 2002. "Basic Principles and Ecological Consequences of Altered Flow Regimes for Aquatic Biodiversity." *Environmental Management* 30 (4): 492–507.

Davis, Richard, and Rafik Hirji, eds. 2003a. "Environmental Flows: Concepts and Methods." Water Resources and Environment Technical Note C1, World Bank, Washington, DC.

———. 2003b. "Environmental Flows: Case Studies." Water Resources and Environment Technical Note C2, World Bank, Washington, DC.

———. 2003c. "Environmental Flows: Flood Flows." Water Resources and Environment Technical Note C3, World Bank, Washington, DC.

———. 2003d. "Water Resources and Environment." Technical Note C1–C3, World Bank, Washington, DC.

Dyson, Megan, Ger Bergkamp, and John Scanlon, eds. 2003. *Flow: The Essentials of Environmental Flows*. Gland, Switzerland, and Cambridge, U.K.: IUCN.

González, Fernando J., Thinus Basson, and Bart Schultz. 2005. "Final Report of IPOE for Review of Studies on Water Escapages below Kotri Barrage." Unpublished manuscript.

Gordon, Nancy D., Thomas A. McMahon, Brian L. Findlayson, Christopher J. Gippel, and Rory J. Nathan. 2004. *Stream Hydrology: An Introduction for Ecologists*. 2d ed. Chichester, U.K.: John Wiley and Sons.

Grey, David, and Claudia Sadoff. 2006. "Water for Growth and Development." In *Thematic Documents of the IV World Water Forum*. Mexico City: Comisión Nacional del Agua.

Hirji, Rafik, and Richard Davis. 2009a. Environmental Flows in Water Resources Policies, Plans, and Projects: Case Studies. Environment Department. Washington, DC: World Bank.

————. 2009b. Strategic Environmental Assessment: Improving Water Resources Governance and Decision Making. Water Sector Board Discussion Paper No. 13. Washington, DC: World Bank.

Hirji, Rafik, Phyllis Johnson, Pail Maro, and Tabeth Matiza Chiuta, eds. 2002. *Defining and Mainstreaming Environmental Sustainability in Water Resources Management in Southern Africa*. Maseru, Lesotho: Southern Africa Development Community, IUCN, Southern Africa Research and Documentation Centre, World Bank.

Hirji, Rafik, and Thomas Panella. 2003. "Evolving Policy Reforms and Experiences for Addressing Downstream Impacts in World Bank Water Resources Projects." *River Research and Applications* 19 (5-6): 667–81.

Hirji, Rafik, and Peter L. Watson. 2007. "Environmental Flow Policy Development and Implementation: Lessons from the Lesotho Highlands Water Project." Paper prepared for the International River Symposium, Brisbane, Australia.

Hou, P., R. J. S. Beeton, R. W. Carter, X. G. Dong, and X. Li. 2006. "Responses to Environmental Flows in the Lower Tarim River, Xinjiang, China: Groundwater." *Journal of Environmental Management* 83 (4): 371–82.

IAIA (International Association for Impact Assessment). 2002. *Strategic Environmental Assessment: Performance Criteria*. IAIA Special Publication 1. Fargo, ND: IAIA.

ILEC (International Lake Environment Committee). 2005. *Managing Lakes and Their Basins for Sustainable Use: A Report for Lake Basin Managers and Stakeholders*. Kusatsu, Japan: ILEC.

International Hydropower Association. 2004. *Sustainability Guidelines*. Sutton, U.K.: International Hydropower Association.

IUCN (International Union for the Conservation of Nature). 2000. *Vision for Water and Nature. A World Strategy for Conservation and Sustainable Management of Water Resources in the 21st Century*. Gland, Switzerland: IUCN.

IUCN Vietnam. 2005. *Environmental Flows: Rapid Flow Assessment for the Huong River Basin, Central Vietnam*. Hanoi, Vietnam: IUCN Vietnam.

Jiménez, Jorge A., Julio Calvo, Francisco Pizarro, and Eugenio González. 2005. *Conceptualisation of Environmental Flows in Costa Rica. Preliminary Determination for the Tempisque River*. San José, Costa Rica: IUCN.

Kansiime, Frank, and Maimuna Nalubega. 1999. *Wastewater Treatment by a Natural Wetland: The Navivubo Swamp, Uganda; Processes and Implications*. Ph.D. thesis, Wageningen Agricultural University, Wageningen, the Netherlands.

Keeney, Ralph. 1992. *Value-Focused Thinking: A Path to Creative Decisionmaking*. Cambridge, MA: Harvard University Press.

King, Jackie, Cate Brown, and Hossein Sabet. 2003. "A Scenario-Based Holistic Approach to Environmental Flow Assessment for Rivers." *River Research and Applications* 19 (5-6): 619–39.

King, Jackie M., and Rebecca E. Tharme. 1994. *Assessment of the Instream Flow Incremental Methodology and Initial Development of Alternative Methodologies for South Africa*. Water Research Commission Report 295/94. Pretoria, South Africa: Water Research Commission.

King, Jackie M., Rebecca E. Tharme, and M. S. de Villiers, eds. 2000. *Environmental Flow Assessments for Rivers: Manual for the Building Block Methodology*. Water Research Commission Report TT 131/00. Pretoria, South Africa: Water Research Commission.

Klasen, Stephan. 2002. *The Costs and Benefits of Change Requirements (IFR) below the Phase 1 Structures of the Lesotho Highlands Water Project (LHWP)*. Maseru: Lesotho Highlands Development Authority.

Ledec, George, and Juan David Quintero. 2003. *Good Dams and Bad Dams: Environmental Criteria for Site Selection of Hydroelectric Projects*. Washington, DC: World Bank.

Lesotho Highlands Development Authority. 2007. *Instream Flow Requirements Audit for Phase 1 Dams of the Lesotho Highlands Water Project*. Maseru, Lesotho: Lesotho Highlands Development Authority.

Millennium Ecosystem Assessment. 2005. *Ecosystems and Human Well-Being: Biodiversity Synthesis*. Washington, DC: World Resources Institute.

Ministry of Planning and Development, Trinidad and Tobago. 1999. *Water Resources Management Strategy for Trinidad and Tobago: Final Report, Main Report*. Port of Spain: Government of Trinidad and Tobago.

Ministry of Water and Livestock Development, Tanzania. 2002. *Project Report: The Sustainable Management of the Usangu Wetland and Its Catchment; December 1998–March 2002*. Dar es Salaam: Ministry of Water and Livestock Development.

Mogaka, Herzon, Samuel Gichere, Richard Davis, and Rafik Hirji. 2004. *Impacts and Costs of Climate Variability and Water Resources Degradation in Kenya: Rationale for Promoting Improved Water Resources Development and Management*. Washington, DC: World Bank.

Murray-Darling Basin Commission. 2000. *Review of the Operation of the Cap: Overview Report of the Murray-Darling Basin Commission*. Canberra, Australia: Murray-Darling Basin Commission.

National Water Commission. 2007. *National Water Initiative: First Biennial Assessment of Progress in Implementation*. Canberra, Australia: National Water Commission.

Nature Conservancy. 2006. *Environmental Flows: Water for People, Water for Nature*. TNC MRCSO1730. Boulder, CO: Nature Conservancy.

Nature Conservancy and Natural Heritage Institute. Forthcoming. "Integrating Environmental Flows in Hydropower Dam Planning, Design, and Operations." Technical Guidance Note. Washington, DC: World Bank.

Ortolano, Leonard, Brian Jenkins, and Ramon P. Abracosa. 1987. "Speculations on When and Why EIA Is Effective." *Environmental Impact Assessment Review* 7 (4): 285–92.

Palmieri, Alessandro, Farhed Shah, George Annandale, and Ariel Dinar. 2003. *Reservoir Conservation: Economic and Engineering Evaluation of Alternative Strategies for Managing Sedimentation in Storage Reservoirs*. Vol. 1: *The RESCON Approach*. Washington, DC: World Bank.

Postel, Sandra, and Brian Richter. 2003. *Rivers of Life: Managing Water for People and Nature*. Washington, DC: Island Press.

Richter, Brian D., Jeffrey V. Baumgartner, Jennifer Powell, and David P. Braun. 1996. "A Method for Assessing Hydrological Alteration within Ecosystems." *Conservation Biology* 10 (4): 1163–74.

Roderick, Michael L., Leon D. Rotstayn, Graham D. Farquhar, and Michael T. Hobbins. 2007. "On the Attribution of Changing Pan Evaporation." *Geophysical Research Letters* 34 (17): L1740.

Scanlon, John. 2006. "A Hundred Years of Negotiations with No End in Sight: Where Is the Murray-Darling Basin Initiative Leading Us?" Keynote address, Environment Institute of Australia and New Zealand Conference, Adelaide, South Australia, September.

Scanlon, John, Angela Cassar, and Noémi Nemes. 2004. *Water as a Human Right?* Gland, Switzerland, and Cambridge, U.K.: IUCN.

Smakhtin, Vladimir U., and Markandu Anputhas. 2006. "An Assessment of Environmental Flow Requirements of Indian River Basins." Research Report 107. Colombo, Sri Lanka: International Water Management Institute.

Smakhtin, Vladimir U., R. L. Shilpakar, and D. A. Hughes. 2006. "Hydrology-Based Assessment of Environmental Flows: An Example from Nepal." *Hydrological Sciences Journal* 51 (2): 207–22.

Smakhtin, Vladimir U., and Neelanga Weragala. 2005. "An Assessment of the Hydrology and Environmental Flows in the Walawe River Basin, Sri Lanka." Research Report 103, Colombo, Sri Lanka: International Water Management Institute.

Tharme, Rebecca E. 2003. "A Global Perspective on Environmental Flow Assessment: Emerging Trends in the Development and Application of Environmental Flow Methodologies for Rivers." *Rivers Research and Applications* 19 (5-6): 397–441.

Tharme, Rebecca E., and Jackie M. King. 1998. "Development of the Building Block Methodology for Instream Flow Assessments and Supporting Research on the Efforts of Different Magnitude Flows on Riverine Ecosystems." Water Research Commission Report 576/198, Pretoria, South Africa: Water Research Commission.

van Wyk, Ernita, Charles M. Breen, Dirk J. Roux, Kevin H. Rogers, T. Sherwill, and Brian W. van Wilgen. 2006. "The Ecological Reserve: Towards a Common Understanding for River Management in South Africa." *Water South Africa* 32 (3): 403–09.

Walker, Keith, Fran Sheldon, and James T. Puckridge. 1995. "A Perspective on Dryland River Ecosystems." *Regulated Rivers* 11 (1): 85–104.

Watson, Peter L. Forthcoming. *Managing the River as Well as the Dam: Designing and Implementing an Environmental Flow Policy; Lessons Learned from the Lesotho Highlands Water Project.* Washington, DC: World Bank.

World Bank. 1991. *Environmental Assessment Sourcebook.* Washington, DC: World Bank.

———. 1993. *Water Resources Management Policy.* Washington, DC: World Bank.

———. 1998. "Project Appraisal Document: Tarim Basin II Project, China." P046563, World Bank, Washington, DC.

———. 2001a. "Improving Performance in Water Management: Bank–Netherlands Water Partnership Program Project Brief." World Bank, Washington, DC.

———. 2001b. *Making Sustainable Commitments: An Environment Strategy for the World Bank.* Washington, DC: World Bank.

———. 2006a. "Project Appraisal Document. Senegal River Basin Water Resources Development Project." World Bank, Washington, DC.

———. 2006b. *Reengaging in Agricultural Water Management: Challenges and Options.* Washington, DC: World Bank.

———. 2006c. "Tanzania Water Resources Assistance Strategy: Improving Water Security for Sustaining Livelihoods and Growth." Report 35327-TZ, World Bank, Washington, DC.

World Commission on Dams. 2000. *Dams and Development: A New Framework for Decision-Making.* London and Sterling, VA: Earthscan Publications.

World Wide Fund for Nature. 2000. *Living Planet Report 2000.* Gland, Switzerland: World Wide Fund for Nature.

Young, William J. 2004. "Water Allocation and Environmental Flows in Lake Basin Management." Thematic paper presented to Lake Basin Management Initiative, International Lake Environment Management Committee, Kusatsu, Japan.

Zhang, Lu, Warwick Dawes, and Glen Walker. 1999. "Predicting the Effect of Vegetation Changes on Catchment Average Water Balance." Technical Report 99/12. Canberra, Australia: Cooperative Research Centre for Catchment Hydrology.

I N D E X

Boxes, figures, notes, and tables are indicated by b, f, n, and t, respectively.

LaVergne, TN USA
23 February 2010
173954LV00008B/76/P